How James Watt Invented the Copier

René Schils

How James Watt Invented the Copier

Forgotten Inventions of Our Great Scientists

Translated by Andy Brown

 Springer

René Schils
Haneweg 25
8181 RZ Heerde
The Netherlands
rene.schils@hotmail.nl

ISBN 978-1-4614-0859-8   e-ISBN 978-1-4614-0860-4
DOI 10.1007/978-1-4614-0860-4
Springer New York Dordrecht Heidelberg London

Library of Congress Control Number: 2011941218

Translation from the Dutch language edition: Einsteins koelkast: en andere vergeten ontdekkingen van grote wetenschappers by René Schils, © Veen Magazines 2008. All rights reserved
© Springer Science+Business Media, LLC 2012

Printed on acid-free paper

Springer is part of Springer Science+Business Media (www.springer.com)

Contents

Johannes Kepler

The astronomer Johannes Kepler is most well known for his three laws describing the motion of planets around the Sun. Symmetry was an overriding principle in Kepler's work, not only at the macroscopic scale of the universe but also at the microscopic scale of a snowflake. He was the first to try to explain the sixfold symmetry of snow crystals.

R. Schils, *How James Watt Invented the Copier: Forgotten Inventions of Our Great Scientists*, DOI 10.1007/978-1-4614-0860-4_1,

Kepler's Laws

While studying at the German university of Tübingen, Johannes Kepler became acquainted with Copernican astronomy, which claimed that the Sun, rather than the Earth, was at the center of the solar system. His talents were soon recognized and, at the early age of 23, he was teaching mathematics at a Lutheran high school in Graz. During his stay in Graz, Kepler published *Mysterium Cosmograficum*, his first great work on astronomy, in which he described a model of the solar system based on the five Platonic solids. These are regular polygons with 4, 6, 8, 12, and 20 faces, known as the tetrahedron, cube, octahedron, dodecahedron, and icosahedron, respectively. Kepler was convinced that this symmetry revealed God's geometric plan for structuring the universe. The orbits of the planets pass over the spheres that fit exactly within or around the regular polygons. His model produced remarkably good results, except for the innermost planet, Mercury. Notwithstanding this magnificent model, the dimensions of the planetary orbits are today considered to be random.

In 1600, as the assistant of Tycho Brahe, Kepler had the opportunity to conduct new calculations of the planetary orbits, this time using Brahe's observations, which were very accurate for the time. Their collaboration unfortunately came to an abrupt end in 1601, when Brahe died unexpectedly. Two days later, Kepler was appointed his successor and was entitled to call himself the "imperial mathematician." In 1609, Kepler published his most famous work, *Astronomia Nova*, which contained his first two laws. What started as an analysis of the orbit of Mars, eventually led to a general description applicable to the orbits of all the then known planets. Kepler's first law states that planets follow an elliptical orbit, with the Sun at one of the two foci. His second law says that a line joining a planet and the Sun sweeps out equal areas during equal intervals of time. That means that the closer a planet is to the Sun, the greater its velocity. Kepler published his third law, which describes the relationship between planets' orbital periods and their distance from the Sun, 10 years later in his *Harmonices Mundi*.

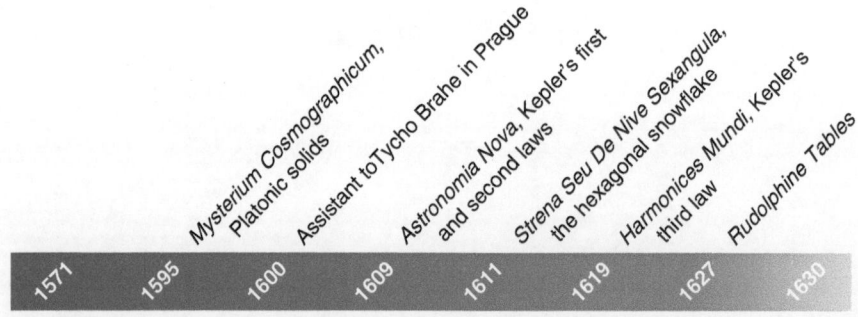

Kepler spent the remaining years of his life fulfilling Tycho Brahe's wish to produce an astronomical table for Emperor Rudolf II. Using Brahe's data, he published the Rudolphine Tables in 1627, an extensive catalogue describing the positions of more than 1,400 stars.

Snow Crystals

In the winter of 1610, Johannes Kepler was looking through the window of his home in Prague at the whirling snow outside. The wintery scene inspired him to write a small book on the hexagonal snowflake, *Strena Seu de Nive Sexangula*, as a New Year's present to his friend and patron, Johannes Matthäus Wackher von Wackenfels. In that time, it was customary to give each other presents on the first day of the New Year. In the book, Kepler explored why snowflakes are hexagonal. It was the first scientific study that tried to explain the structure of snow crystals. The hexagonal structure of snowflakes had been documented long before: the first texts date back to ancient China where Han Ying wrote as early as 135 BC that, while the flowers of trees and plants are generally pentagonal, snow is always hexagonal. More than twelve centuries later, the Chinese philosopher Chu Hsi noted that six is the perfect number for water, explaining why, when snow turns into crystal flowers, they always have six sides.

Kepler was never satisfied with knowing how things fitted together, but always wanted to know why. He applied his ideas to the immense scale of the universe and to the small scale of matter. At both the macro- and microscale he found answers in geometry and symmetry or, as he himself said: "Where there is matter, there is geometry." Inspired by examples from nature, Kepler wrote that "the six sided shape of a snowflake is none other than that of the ordered shapes of plants." Kepler was very religious and did not believe that the ordered pattern of a snowflake occurred at random. From this basic principle, he developed a framework on the structure of matter that bears an amazing resemblance to current thinking on crystal structures. And he did that 200 years before Dalton devised his theory of the atom as the basic unit of all matter.

To explain the hexagonal shape of snowflakes, Kepler went in search of other hexagons that occur in nature, such as a honeycomb or the seeds of a pomegranate. The architecture of a honeycomb is such that each cell shares six walls with the neighboring cells in the same row. But it fascinated Kepler even more that the cells in one row are joined to those in the opposing row by three diamond-shaped faces. In this way, he made the step from the perfect combination of six-cornered faces in two dimensions to three-dimensional regular polygons. Kepler then elaborated that this geometric symmetry can be explained by the most efficient manner of filling up a space. It is similar to the pattern you get if you press a large number of small spheres into a round barrel.

So Kepler set about stacking spheres. He stacked them in different ways, trying to find the most efficient way, with the smallest possible space between them.

He achieved the greatest density by stacking the spheres in the empty spaces in the underlying layer, in the same way that a greengrocer will stack oranges. Although he could not explain why, he believed this was the most efficient method of packing: "It is the tightest possible; in no other arrangement can more spheres be accommodated in the same vessel." A present-day crystallographer would call this *face-centered cubic* packing with a density of 74.05%. Mathematicians have wracked their brains for three centuries trying to prove the "Kepler conjecture." To prove that this is the most efficient form of packing, all other possibilities have to be calculated

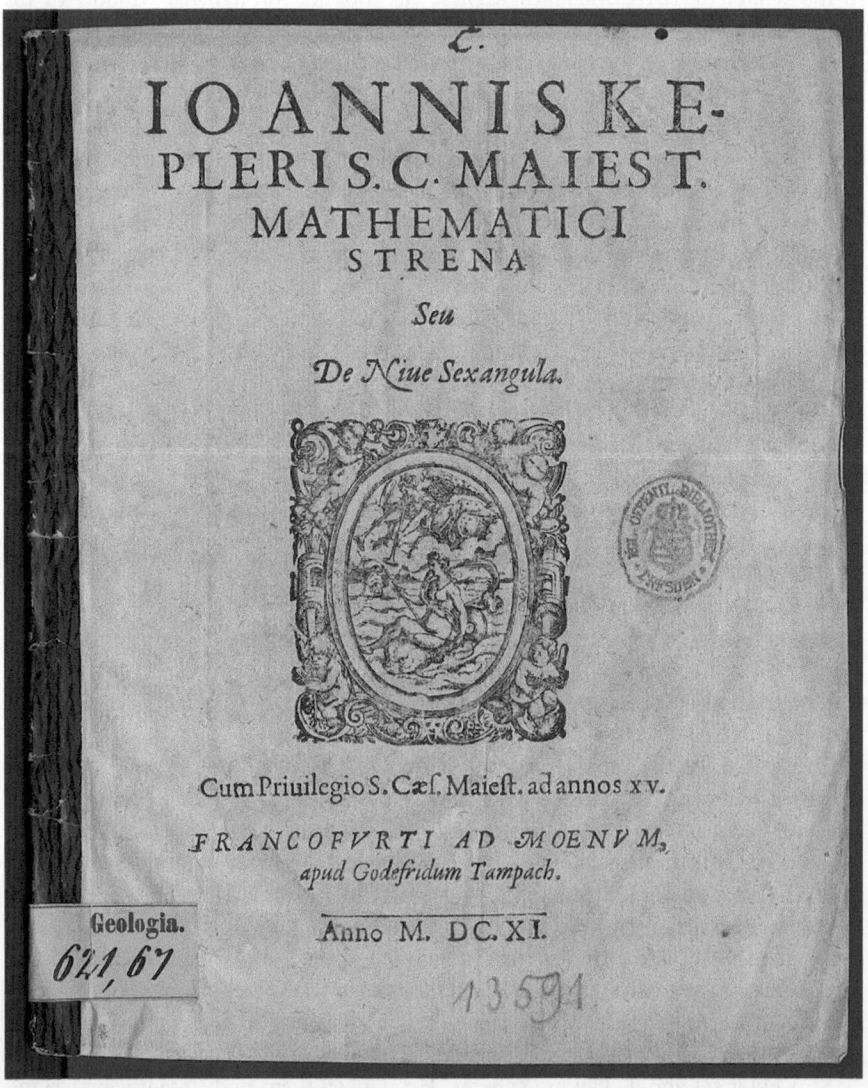

The cover of the 15-page booklet by Johannes Kepler

The American Wilson Bentley took the first photographs of snowflakes in 1885

as well, including the irregular arrangements. In 1965, it was demonstrated that "only" 5,000 packing arrangements were eligible for the first prize. But it was not until 1998 that mathematician Thomas Hales was eventually able to prove that Kepler was right.

Unfortunately, Kepler never completely unraveled the secrets of snow. He just could not prove why snow crystallizes in a hexagonal – rather than a triangular or cubic – shape. He suspected that there must be some kind of design principle, but could not find it himself. That question he left to the chemists. Kepler had laid the foundations for the scientific study of snow crystals, but then left the initiative to others.

In Europe, one of those who continued Kepler's work was René Descartes, who once again confirmed the link between the six-sided symmetry and the uniform packing of crystals. He succeeded in providing a remarkably accurate description of snow crystals through observation with his naked eye. In the seventeenth century, the first microscopes were developed. Robert Hooke used a microscope to study the morphology of snowflakes, and discovered the complex structure that is typical of the featherlike pattern of a snowflake. The first photographs of an individual snow- flake were taken by Wilson Bentley in 1885. Using a self-made camera, he photo- graphed around 5,000 different snowflakes, showing once again that "no two snowflakes are the same." Descartes and Bentley were so impressed by the unique symmetry of a snowflake that they concluded that such a beautiful shape could never be created by man. In the 1950s, however, Ukichiro Nakaya succeeded in making artificial snow crystals.

Snowflakes actually consist of smaller snow crystals, which in turn consist of a conglomeration of ice crystals. The ice crystals cluster together to make snow crys- tals around small cores of dust or salt in the air. The snow crystals then accumulate to form snowflakes, whose ultimate shape and properties depend on temperature, air humidity, and wind velocity. Nakaya's research provided a wealth of new knowl- edge on the properties of different types of snow, which scientists now make grateful use of to assess the risks of avalanche.

Around 300 years after Kepler, X-ray crystallography revealed the atomic struc- ture of ice crystals. When water freezes, it forms a hexagonal network of water mol- ecules. The molecules are linked to each other by hydrogen bridges, but are certainly not densely stacked. Finally, Kepler's question has been answered.

References

Johannes Kepler, 1966. *The Six-Cornered Snowflake*. Clarendon Press, 75 pp. English Translation
    of the original *Strena Seu De Nive Sexangula* from 1611.
Cecil Schneer, 1960. 'Kepler's New Year's Gift of a Snowflake'. *Isis* 51 (4), 531–545.
Omolar Olowoyeye, 2003. 'The History of the Science of Snowflakes'. *Dartmouth Undergraduate
    Journal of Science* 5 (3), 18–20.

Robert Hooke

We know Robert Hooke mainly from the law that bears his name, which describes the extension of a spring as a function of the force applied to it. Other than that, Hooke has been almost forgotten. Undeservedly, as this "English Da Vinci" was a great and many-faceted scientist. After the Great Fire of London, he played a prominent role in the reconstruction of the city, not least as an architect.

R. Schils, *How James Watt Invented the Copier: Forgotten Inventions of Our Great Scientists*, DOI 10.1007/978-1-4614-0860-4_2,
© Springer Science+Business Media, LLC 2012

Hooke's Law

It is an experiment that almost all of us have conducted at least once at school: hang a series of different weights on a spring and measure the different degrees to which it extends. Hooke's Law states that the extension of a spring is in direct proportion with the load applied to it. Hooke published his law in 1679 in *De Potentia Bestitutiva*. It was just a small section in a comprehensive analysis of vibration and elasticity.

Three years earlier he lifted a corner of the veil in the form of an anagram, a popular way of making a discovery known in the seventeenth century. In *Helioscopes* Hooke announced his law as "cediinnoopsssttuu," an anagram of "Ut Pondus sic Tensio" (As the extension, so the weight). Hooke used the term "weight" for what would later be called "force" as, before Isaac Newton, the two concepts had not been clearly distinguished.

Hooke was a contemporary of renowned scientists like Newton, Robert Boyle, and Edmond Halley, who he met during his studies at Oxford and as Curator of Experiments at the Royal Society. Hooke's scientific interest varied from the small world of insects to the large world of the planets. In his book *Micrographia*, Hooke described the world through the lens of microscope and telescope. He is famous, for example, for his drawing of a flea but was equally at home observing the surface of the Moon and Jupiter.

During his life, Hooke regularly clashed with other scientists. He corresponded regularly with Newton on gravity, but when the latter published his theories, Hooke felt that he had not been given sufficient credit for his contribution. He had a similar conflict with Christiaan Huygens about who had invented the balance spring, an essential component of clocks and watches.

Despite Hooke's great scientific achievements, his name has been forced into the background over time and he has been overshadowed by his famous contemporaries.

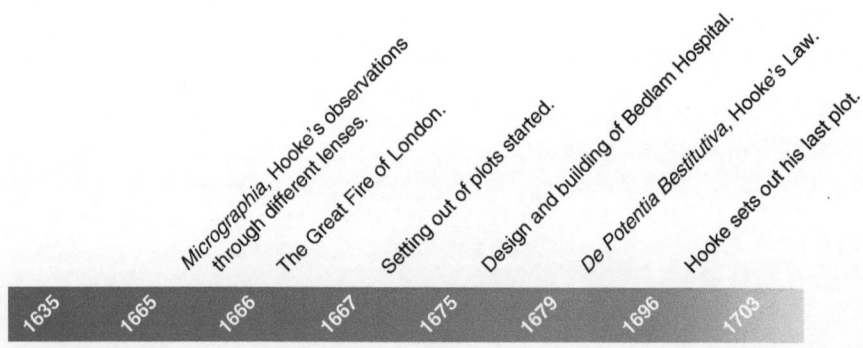

Robert Hooke

Typical of Hooke's banishment from scientific history is the lack of any pictures of him. According to reports by visitors to the Royal Society, there was still a portrait of Hooke next to that of Robert Boyle in 1710, but after that it disappeared. Newton is alleged to be responsible for the disappearance. Later, a stained glass window bearing a figurative portrait of Hooke was placed in St. Helen's Church in Bishopsgate, London. That, too, was short-lived, as the church was badly damaged in 1993 when an IRA bomb exploded in the financial heart of the city.

Land Surveyor and Architect

In the early morning of Sunday September 2, 1666, a fire broke out in a bakery in Pudding Lane. Five days later, the "Great Fire" had destroyed more than 13,000 houses and made 65,000 people homeless. Of the city's 109 churches, 87 were burned out, including St. Paul's cathedral – the greatest disaster of all. Only 20% of the area within the city walls had emerged unscathed. These were fearful days for Robert Hooke. The powerful wind spread the fire so quickly that he was afraid that it would destroy his home in Gresham College, but fortunately the flames came to a

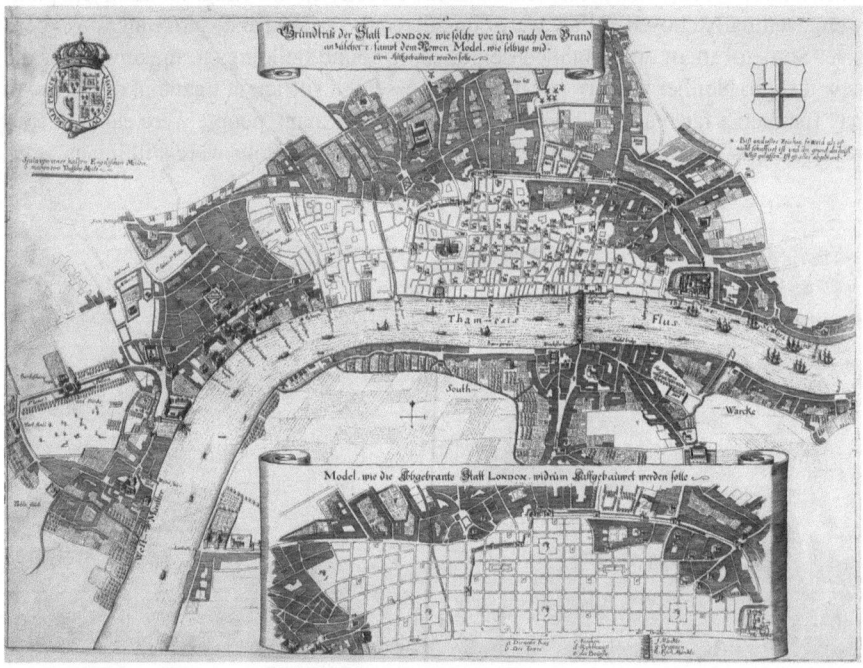

Map of London after the fire, with the burned out area shown in white. The insert at the bottom may be Robert Hooke's street plan

halt just before they reached his block. Only a few steps from his door, Hooke found himself in the middle of the smoldering ruins of his city.

On the Thursday that the fire finally burned itself out completely, London's civic leaders met in Gresham College to discuss how to recover from the crisis. Only a charred skeleton remained of the original city hall, so Gresham College rapidly became the center of municipal government. For Hooke it heralded the start of a career he could never have imagined. Without asking for it, he suddenly found himself right at the spot where important decisions were being made about the reconstruction of London.

In the years that followed, Hooke's time was almost completely taken up by the reconstruction, first as a land surveyor and then as an architect. In the eyes of the city's leaders, his familiarity with the local area and his knowledge of geometry made him indispensable. For someone who had never had anything to do with running a city, Hooke enjoyed a considerable degree of trust.

After the fire had been extinguished, the people of London immediately started rebuilding their houses. King Charles II found this a little too hasty and instructed the city leaders to order all building activities to be halted until it was clear how the city should be reconstructed. On September 21, Hooke presented his plans to the city council. It consisted of a radical grid-style plan, with streets running only north–south and east–west, similar to the design of many modern American cities. Besides Hooke's proposal, a number of other plans had been submitted, including one by Christopher Wren, who presented his ideas directly to the King. None of the plans were eventually chosen as they were considered too radical and, above all, too expensive. Some of them even entailed demolishing sound buildings to make way for the new design. Neither the city nor the King possessed sufficient financial resources to pay for such a far-reaching operation. Furthermore, many people were eager to start rebuilding as, despite the great destruction, many foundations were still intact.

Bedlam Hospital, designed and built by Robert Hooke

On October 4, 1666, the King appointed Wren as the Royal representative in the official rebuilding commission. In turn, the city nominated Hooke as its representative. The most idealistic plans had already been dismissed and had given way to the practical realities of determining the fire damage and clearing the rubble. The rebuilding commission specified which streets should be widened and which alleyways should disappear. In the commission, Hooke was very busy drawing up new building specifications. The new houses had to be built of stone or brick and had to be cleaner, healthier, and safer than the old ones.

Six months after the fire, together with three others, Hooke was appointed by law as official surveyor for the reconstruction of London. On March 27, 1667, a start was made on setting out the streets, beginning with Fleet Street. After 9 weeks, they had set out the main streets, but it eventually took 2 years to complete the job.

In the meantime, they were also able to start setting out the plots for building. The first plot was designated on May 13, 1667. Nearly 30 years later, they had set out almost 8,400 plots, some 3,000 of which had been done by Hooke. House owners could submit a request to the city council for their plot to be designated. After the owner had cleared away all the rubble, the surveyor would visit the site and try to redesignate the plot on the basis of the old foundations. If necessary, he would use additional information provided by the owner himself or neighbors.

Generally, there were few problems designating the new plots and they encountered little resistance from the owners. Of course, there were disputes between neighbors and some people were unhappy as they had to give up part of their plots to widen streets or enlarge public buildings. Money was made available for compensation, raised through a tax on coal. The city council determined the level of compensation, usually on Hooke's advice.

Much has been written about the cooperation and power struggle between Wren and Hooke. In the eyes of the city, Hooke was the hero, who took the initiative at the right moment. For the Crown, however, it was Wren who was the hero: the young, visionary architect who was in the right place at the right time. The fact is that Wren was the prominent figure, made responsible by the King for rebuilding all public buildings and churches, including of course St. Paul's. He and Hooke, however, had an excellent relationship and they worked closely together, consulting each other almost daily. The more the construction plan for London took shape, the more Wren involved Hooke in the architectural work. In December 1670, Hooke was given his first independent assignment: to rebuild the new Royal College of Physicians, with an anatomical theater modeled on the one at the University of Leiden. This was to be followed by others, especially for private clients.

Around 1675, Hooke designed and built Bedlam Hospital, intended as a home for mental patients. This time he used the Palais des Tuileries in Paris as an example, much to the displeasure of Louis XIV, who considered it a downright insult that his palace should serve as an example for a lunatic asylum.

Hooke was above all a good technical draftsman, but never became a great architect. He had a good feeling for proportion and his buildings were pleasing to the eye. He was probably inspired by the architecture of the Doric order, one of the three architectural orders of classical Greece, which is characterized by very stately, clean lines.

The only one of Hooke's buildings that still stands is the monument to the Great Fire, a stately column 70 m high, topped off with a gilded urn of fire. Because of the close cooperation between Wren and Hooke, it is not certain who was responsible for the design of many of the buildings. Many of Hooke's drawings later turned up in an overview of Wren's designs and were therefore erroneously attributed to the latter.

During the rebuilding of London, Hooke had to deal with a large number of technical and organizational questions. For example, he devoted himself to finding the best way to build an arch. He discovered that the line of an arch that has to support a certain weight must be the inversion of a catenary, or free-hanging chain, with the same weight. In an appendix to *Helioscopes* he wrote, again in an anagram, that he had found "a true mathematical and mechanical form of all manner of Arches for Building." Two years after his death, his executor revealed the meaning of the anagram: "Ut pendet continuum flexile, sic stabit contiguum rigidum inversum" (As hangs a flexible cable so, inverted, stand the touching pieces of an arch).

References

Henry William Robinson, 1948. 'Robert Hooke as a Surveyor and Architect'. *Notes and Records of the Royal Society of London* 6 (1), 48–55.

Michael Cooper, 1997. 'Robert Hooke's Work as Surveyor for the City of London in the Aftermath of the Great Fire'. *Notes and Records of the Royal Society of London* 51 (2) 161–174 (part 1), 52 (1) 25–38 (part 2), 52 (2) 205–220 (part 3).

Lisa Jardine, 2003. *The Curious Life of Robert Hooke, the Man Who Measured London.* HarperCollins Publishers, 422 pp.

Edmond Halley

Edmond Halley's name is forever
associated with the comet that passes
close by the Earth every 75 years. In
1705, more than half a century in
advance, Halley correctly predicted
the year in which the comet would
return. Of a more worldly nature was
his life table, which is still considered
a milestone in actuarial science.

R. Schils, *How James Watt Invented the Copier: Forgotten Inventions*
of Our Great Scientists, DOI 10.1007/978-1-4614-0860-4_3,
© Springer Science+Business Media, LLC 2012

Comet

At Oxford, Edmond Halley became acquainted with the highly respected scientist John Flamsteed, England's first Astronomer Royal. Halley visited Flamsteed on several occasions at the newly established observatory at Greenwich. Inspired by Flamsteed's catalogue of the stars in the northern hemisphere, Halley traveled to the island of St. Helena to catalogue the southern night sky. From the most southerly point of the British Empire, Halley determined the positions of 341 stars. During his stay of more than a year he also witnessed the passage of Mercury between the Sun and the Earth.

Back in London, he soon became a member of the scientific elite, despite having left Oxford without a degree. To rectify that omission, he was awarded a degree by Royal decree without having to take any examinations. Together with contemporaries like Robert Hooke and Isaac Newton, he sought a mechanical explanation for the orbits of the planets. Where Hooke and Halley failed in their endeavors, Newton eventually succeeded. Newton, however, lacked the perseverance to put his findings on paper. It was only due to Halley's psychological and financial support that, in 1687, he finally published one of the greatest scientific classics of all time, *Philosophiae Naturalis Principia Mathematica*. Since the book would probably never have seen the light of day without Halley's support, he has been called the midwife of Newton's *Principia*.

In 1704, Halley was appointed Savilian professor of geometry at the University of Oxford, but that by no means signified the end of his studies in astronomy. On the contrary, within a year he had written *A Synopsis of the Astronomy of Comets*,

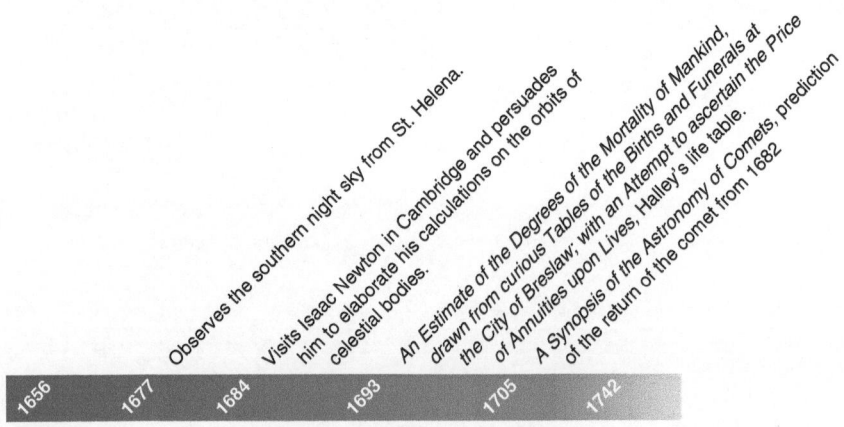

Edmond Halley

in which he described the parabolic orbit of 24 comets observed between 1337 and 1698. For these 24 comets he found only 22 different orbits. Three comets seemed to be following the same orbit: the comets of 1531, observed by the German Apianus, 1607, observed by none other than Johannes Kepler, and lastly of 1682, whose orbit had been accurately described by Flamsteed.

Halley concluded that these must be one and the same comet, and predicted accurately that it would return in 1758. He himself did not live to see his prediction come true, but since then the comet has of course borne his name. Halley's Comet passes faithfully by the Earth every 75 or 76 years, most recently in 1986.

The Price of a Life Annuity

No one can predict when they are going to die. For each of us, the duration of our lives is determined by a variety of factors, including gender, lifestyle, health, hereditary characteristics, and calamity, war or other external events. People have always been aware of the risk of death or of loss of income. The Greeks and Romans had simple systems to pay for pensions and funerals. From the seventeenth century, however, the concept of risk was addressed in a more scientific way, with the realization that life expectancy could be predicted if applied to a large group of people.

Toward the end of the seventeenth century, Edmond Halley was employed by the leading scientific institution in England, the Royal Society. Although, until then, he had largely been active as an astronomer, the Royal Society was the ideal place to reflect on a wide range of scientific topics. At some point, he found a document on his desk containing records of all registered births and deaths in the German town of Breslaw over a 5-year period.

To this day, it remains unclear whether Halley simply received these figures or had requested them. Either way, he considered them of great value and used the unique data to draw up a "life table," showing the probability of someone of a certain age dying during that year. He published the life table in an article entitled "An Estimate of the Degrees of the Mortality of Mankind, drawn from curious Tables of the Births and Funerals at the City of Breslaw; with an Attempt to ascertain the Price of Annuities upon Lives." As the title suggests, he also used the life table to calculate life annuities. This assured him a permanent place in the history of actuarial science.

When Halley performed his calculations, he was well acquainted with the earlier work of John Graunt and William Pett, who drew up much less accurate life tables in 1661 using figures from the "Bills of Mortality" of the cities of London and Dublin. At the time, the Bills of Mortality were the major source of statistics on causes of death. They were originally introduced as a kind of early warning system for imminent outbreaks of the much-feared plague and therefore focused on the cause of death, initially failing to record the age of the deceased. Graunt and Petty solved this problem by estimating the age as closely as possible on the basis of the cause of death. Certain diseases were known, for example, only to occur in children. Graunt and Petty were also forced to make a rough estimate of the total population,

7 . 8 9 .   .   14   . 18 . 21 . 27 . 28 . . 35 .
11 . 11 . 6 . $5\frac{1}{2}$. 2 . $3\frac{1}{2}$ 5 6 $4\frac{1}{2}$ $6\frac{1}{2}$ 9 . 8 . 7 . 7 .

36 .   42 .   45   49 54 .   55 . 56   . 63
8 . $9\frac{1}{2}$ 8 . 9 . 7 . 7   . 10 11 .   9 .   9 . 10 . 12

    70 71 . 72   77   81   84 .   90 91   .
$9\frac{1}{2}$ 14 9   . 11 $9\frac{1}{2}$ 6 . 7 . 3 . 4 . 2 . 1 . 1 . 1   .

98 . 99 . 100.
0 . $\frac{1}{5}$. $\frac{3}{5}$

Age. Curt.	Per- sons.	Age. Curt.	Per- sons.	Age. Curt.	Per- sons.	Age. Curt.	Per- sons.	Age. Curt.	Per- sons.	Age. Curt.	Per- sons.	Age.	Persons.
1	1000	8	680	15	628	22	586	29	539	36	581	7	5547
2	855	9	670	16	622	23	579	30	531	37	472	14	4584
3	798	10	661	17	616	24	573	31	523	38	463	21	4279
4	760	11	653	18	610	25	567	32	515	39	454	28	3964
5	732	12	646	19	604	26	560	33	507	40	445	35	3604
6	710	13	640	20	598	27	553	34	499	41	436	42	3178
7	692	14	634	21	592	28	546	35	490	42	427	49	2709
Age. Curt.	Per- sons.	Age. Curt.	Per- sons.	Age. Curt.	Per- sons.	Age. Curt.	Per- sons.	Age. Curt.	Per- sons.	Age. Curt.	Per- sons.	56	2194
43	417	50	346	57	272	64	202	71	131	78	58	63	1694
44	407	51	335	58	262	65	192	72	120	79	40	70	1204
45	397	52	324	59	252	66	182	73	109	80	41	77	692
46	377	53	313	60	242	67	172	74	98	81	34	84	253
47	377	54	302	61	232	68	162	75	88	82	28	100	107
48	367	55	292	62	222	69	152	76	78	83	23		34000
49	357	56	282	63	212	70	142	77	68	84	20		Sum Total.

Original presentation of the mortality data for Breslaw, as processed by Halley. The most important figures are in the table with a column for age (Age Curt.) and the number of living persons of that age (Persons.). From these figures, Halley deduced the mortality rates shown above the table in two rows; the top row is the age and the lower is the probability of death. Lastly, to the right of the central table, he has determined the total population of Breslaw by adding together the number of people in each 7-year age group

as there were no reliable figures for that either. This was rendered more difficult by the fact that migration made the population very unstable. In the large cities of London and Dublin, many people died who had not been born there, resulting in a large difference between the numbers of births and deaths. Despite these methodological restrictions, the work of Graunt and Petty showed that mortality patterns were considerably regular.

In the Netherlands, too, a number of people published life tables before Halley. In 1669, Lodewijck Huygens wrote to his brother Christiaan that he had drawn up a table based on Graunt's table that could help determine life annuities. And in 1671, Johannes de Witt wrote *Waardije van Lyf-renten naer Proportie van Los-renten* ("The Worth of Life Annuities Compared to Redemption Bonds"), commissioned by the States of Holland. De Witt's results were, however, based on assumed mortality rates and were most likely not tested against actual figures.

In the same year, Amsterdam regent Johannes Hudde adopted a different approach. He had access to data on the payments on life annuity contracts sold by the city of Amsterdam in the period from 1586 to 1590. Hudde recorded how long

| Age | Percentage (%) | | |
(Year)	Halley (1685-1690)	Hudde (1587-1672)	Netherlands (recent)
20	5,2	9,4	0,2
25	6,4	9,5	0,3
30	7,8	9,2	0,3
35	9,2	10,2	0,4
40	10,8	11,7	0,8

Comparison of five yearly mortality according to Halley, Hudde and recent data on the Netherlands, derived from Van Ham. Mortality among young adults is now much lower than in the time of Halley and Hudde. The increasing mortality with age in the recent figures is the result of ageing itself, which has now become the most important factor for the higher age groups. In the sixteenth and seventeenth centuries, normal ageing was negligible and death could almost exclusively be attributed to "accidents," such as infectious diseases and war. These accidents affected all age groups equally, as can be seen from Hudde's figures. Halley's figures are, on average, lower than those of Hudde. In Halley's observation period, there was almost no plague in Breslaw, while Hudde's insured subjects were still affected by it. The reason for Halley's mortality rates doubling from 20 to 40 years of age is not clear. One possible cause is that, following the Peace of Münster in 1648, many of Breslaw's residents returned to the countryside, which was now safe again. Later, the city grew again, slowly and steadily. Births in the years after 1648 were therefore lower than would be expected on the basis of Halley's population of 34,000. Halley's observations therefore underestimated mortality among young adults, but this effect decreases in the older age groups

a total of 1,495 people lived after purchasing a life annuity, ordering the data by age on the date of purchase. His handwritten table has been preserved for posterity, as Hudde enclosed it with a letter he wrote to Christiaan Huygens in 1671. He corresponded with both Huygens and De Witt on a more detailed analysis, but did not produce a life table.

The data that Halley used were gathered by Caspar Neumann, a pastor with an interest in science, between 1687 and 1691. Halley acquired them through Henry Justel, the royal librarian. The figures were unique in that they were hardly, if at all, affected by migration. During the 5 years covered by the records, 6,193 people had been born in Breslaw and 5,869 died, a birth surplus of only 65 a year.

Using Neumann's rough figures, Halley drew up a table showing the number of living persons per age group in each year. From this, it was simple to calculate the probabilities of death for each age. Of the 1,000 children 1 year of age, 855 would reach the age of 2, signifying a death risk of almost 15%. In the somewhat older age groups of 14–17, mortality fell to 2–3%. Above that, the risk of death increased again to reach 10% around the age of 70. Although Halley, like Graunt before him, did not know the size of the total population, he estimated it by adding together the number of people in each age group, bringing him to a total of 34,000.

Halley devised a variety of useful applications for the table. Remarkably, the first had nothing at all to do with life annuities, but with warfare. He reasoned that all men between the ages of 18 and 56 were eligible to fight. He suspected that males under 18 were not equipped for the exertions of war or to bear the weight of arms. For men over 56, he foresaw all kinds of problems related to their advanced age.

He concluded that Breslaw had a potential army of 9,027 men, a little more than a quarter of the total population.

After this foray into military science, Halley turned his attention to life annuities. First of all, he demonstrated how the table could be used to calculate that a person of a specific age would have a certain number of years left to live. It showed, for example, that a 40-year-old had a one-in-five chance of dying within 7 years. In a similar way, he reasoned that the table could be used to estimate the life expectancy of an individual of a certain age. He defined life expectancy as the number of years to the age at which the risk of dying was 50%. According to his table, someone of 30 had a further 27–28 years to live.

One of the most innovative elements of Halley's article was the method he developed to determine the price of a life annuity. For each future year, he calculated the amount that would need to be invested now so as to pay out the annuity when the time comes. For each separate year, he multiplied that amount with the probability that a person would still be alive. The sum of all those amounts times the probability of survival was the total annuity.

Halley stressed that such a calculation is a time-consuming exercise. However, as it was the most important application of the life table, it was in his view worth the effort. The annuities to be paid were of course related to the probabilities of survival. The highest annuities would therefore be paid by people in the 10–15 age group, while at 65, the premium would be half that amount.

Both the origins of his life table and his prediction of the return of the comet show that Halley had a great interest in analyzing historical data. Despite the striking difference between these two topics, he succeeded in using historical information to predict future events.

References

Edmond Halley, 1693. 'An Estimate of the Degrees of the Mortality of Mankind, drawn from curious Tables of the Births and Funerals at the City of Breslaw; with an Attempt to ascertain the Price of Annuities upon Lives'. *Philosophical Transactions of the Royal Society of London* 17: 596–610.

M. Greenwood, 1938. 'The First Life Table'. *Notes and Records of the Royal Society of London* 1 (2), 70–72.

Allan Chapman, 1994. 'Edmond Halley's Use of Historical Evidence in the Advancement of Science'. *Notes and Records of the Royal Society of London* 48 (2), 167–191.

David Ipsen, 2004. *Edmond Halley: More than a Man with a Comet*. Xlibris cooperation, 60 pp.

Dirk van Ham, 2005. 'De Tafel van Afsterving van Johannes Hudde'(Johannes Hudde's Life Table). *De Actuaris*, July 2005, 31–33.

Daniel Bernoulli

Daniel Bernoulli is best known for
the physical principle that bears his
name, which states that as a gas or
fluid flows more quickly, the pressure
it exerts will decrease. Economists,
however, commemorate Bernoulli as
the creator of the utility function.

R. Schils, *How James Watt Invented the Copier: Forgotten Inventions
of Our Great Scientists*, DOI 10.1007/978-1-4614-0860-4_4,
© Springer Science+Business Media, LLC 2012

Bernoulli's Principle

Daniel Bernoulli came from the second generation of a family of prominent mathematicians in Basel, Switzerland. Together with illustrious names like Leibniz, Euler, and Lagrange, the Bernoullis dominated mathematics in the seventeenth and eighteenth centuries. In 3 generations, the family supplied no less than 8 world-renowned mathematicians.

Daniel was born in Groningen, where his father was a professor of mathematics at the time. Unfortunately, father Johan and his son did not get on well together. Their lives were dominated by Johan's fear of being overshadowed by his son. He therefore wanted Daniel to study something other than mathematics. But breeding will out, and in 1723 Daniel published *Exercitationes Quaedam Mathematicae*, in which he used mathematics to describe, among other things, the behavior of fluids.

The scientific prestige this brought led him to be appointed to the Russian Academy of Sciences in St. Petersburg, where he settled with his brother Nikolaus. However, fate quickly took a hand and his brother died, leaving Daniel unhappy and lonely. He made plans to return to Basel, but his father arranged for one of his best pupils, Leonhard Euler, to go to St. Petersburg to keep him company. The two became good friends and this contributed to Bernoulli's stay in St. Petersburg becoming the most successful period of his life in scientific terms.

Bernoulli's greatest achievement was undoubtedly the publication of *Hydrodynamica*, in which he describes the relationship between pressure and speed for

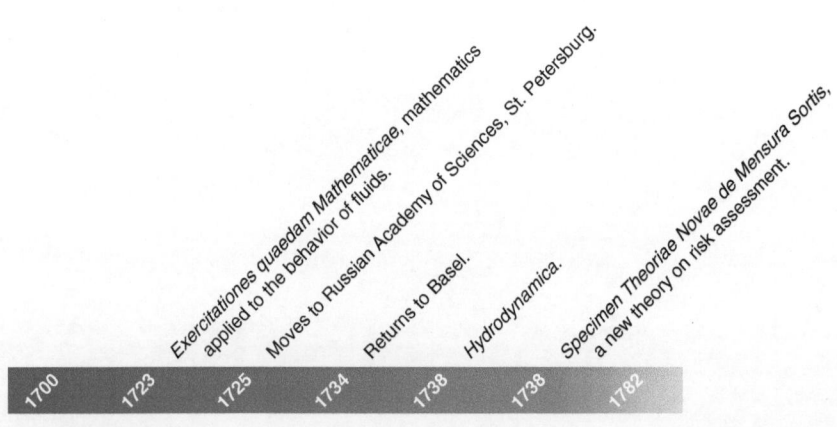

Daniel Bernoulli

fluids. Bernoulli's principle states that as the velocity of a fluid increases, the pressure exerted by that fluid decreases. The principle is a favorite subject for physical practicals, as it can be demonstrated in simple experiments. One way is to hang two paper leaves parallel to each other with a small space between them and blow a little air downwards between them. This will cause them to move together, showing that pressure decreases as the air moves faster.

Bernoulli's principle is best known as an explanation for the aerodynamic lift created by wings. Russian scientist Nikolai Zhukovsky observed that the air flow above a wing was narrower than the flow underneath, meaning that it is also faster than the air passing below the wing. According to Benoulli's principle, pressure is inversely proportional to speed, giving the wing an upward lift. Although this is a very popular example, the reality is a little more complex, as Newtonian laws play a role as well. A wing also remains in the air because of the downward air flow from its tip, caused by the angle of the wing to the direction of movement.

After the success of *Hydrodynamica*, father Bernoulli obviously could not be left behind. He quickly wrote his own work on fluids, entitled *Hydraulica*, using his son's book and, in a final attempt to steal Daniel's laurels, he backdated it to 1732, to give the impression that it had been written first.

Utility Function

From the end of the seventeenth century, Europe was inundated with a tidal wave of lotteries. Besides all kinds of private initiatives, governments regularly organized lotteries to raise funds for bridge-building and other special projects. Generally speaking, the participants were only concerned with winning the first prize and getting rich as soon as possible. They relied primarily on luck, showing little understanding of the underlying probabilities of winning or losing. Daniel Bernoulli's analysis of this popular pastime laid the foundations for one of the cornerstones of current economic theory. He made this diversion into probability theory and economics while he was a professor of botany at the University of Basel. After his successful period in St. Petersburg, his main concern was to return to Basel. The chair for his professorship was only of secondary importance.

In 1738, the same year as *Hydrodynamica* appeared, Bernoulli published his *Specimen Theoriae Novae de Mensura Sortis*, a new theory on how to measure risk. The paper begins with a thought experiment: a poor man finds a lot that offers an equal chance of winning 0 or 20,000 ducats. Would the man be unwise to sell the lot for 9,000 ducats? On the other hand, would a rich man be unwise to buy it for 9,000 ducats? According to Bernoulli, the answer to both questions is "no." Both men therefore rate an identical gamble in a different way. Bernoulli therefore claims that the value of an item should not be determined by the price but by the "utility" it yields. In his view, the thought experiment shows that utility depends on the subjective assessment of the individual concerned.

Daniel Bernoulli used this insight to solve the "St. Petersburg paradox," which had been formulated earlier by his cousin Nicolas Bernoulli. It involves a game in which a coin is tossed repeatedly until it lands on "heads." There is a prize of one ducat if the coin lands on heads after the first throw, and this is doubled with every toss: two ducats for the second toss, four for the third, and so on. The question is, what price would someone be prepared to pay for playing the game? The classical expected value for this game is theoretically infinitely high, namely $\frac{1}{2} \times 1 + \frac{1}{4} \times 2 + 1/8 \times 4 + \ldots + 1/2n \times n = \frac{1}{2} + \frac{1}{2} + \frac{1}{2} + \ldots + \frac{1}{2} =$ infinity. Despite this, people tend in practice to be unwilling to pay much to play. Later, empirical tests showed that players are not usually prepared to pay more than 12 or 13 monetary units. According to Bernoulli, "utility" offers the solution to this paradox. The increase in utility is inversely proportional to the capital that an individual possesses. One ducat will have a higher utility for a beggar than for a banker. A beggar will therefore assess his chances of winning a ducat far differently than a banker. The subjective estimation of the chances of winning is based on calculation of the expected utility. The expected utility of the game is finite as each successive ducat always generates a lower utility than the previous one.

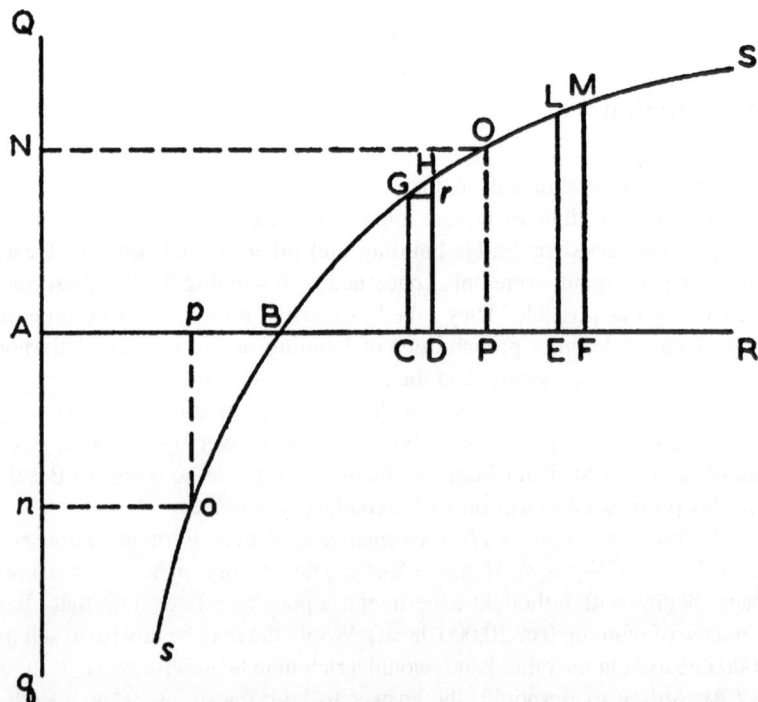

The relationship between utility (Q) on the vertical axis and property (R) on the horizontal axis. The section AB shows property in the original situation. If property increases to P, utility increases from A to N. The same opposing change in utility, from A to n, requires only a decrease in property from B to p. Based on the original figure from Bernoulli's article (1738)

The Dice Shooters' by Dutch artist David Teniers (1610–1690). Interior of an inn with a group of men round a table engrossed in a game of dice. Coins are piled up on the table and in the background a group of farmers are sitting around the fire

The added utility of the high prizes in the St. Petersburg paradox are therefore no longer sufficient to compensate for the lower chances of winning.

It later became clear that utility theory is not at all necessary to solve the St. Petersburg paradox. Strictly speaking, the expected value of the game is only infinitely high if the game is played an infinite number of times. If it is played only once or a few times, the expected value is much lower.

Despite this misconception, Bernoulli's article contains two ideas that have had a significant influence on economic theory. First, the utility of wealth is not linearly related to wealth, but increases in decreasing steps: this is the notion of "diminishing returns." Second, people's assessment of risk is based not on the expected value but on the expected utility. In the centuries that followed, Bernoulli's pioneering work was developed further, resulting in the classic book *Theory of Games and Economic Behavior* (1944) by Von Neuman and Morgenstern.

The insurance business also traces its origins back partly to Bernoulli's theory of expected utility when assessing risk or insecurity. The concept of insurance itself dates back to Babylonian times, but did not acquire a mathematical foundation until the seventeenth century. Bernoulli himself worked out an example for Caius, a merchant who wants to ship goods from St. Petersburg to Amsterdam. If the goods arrive safely he can sell them for 10,000 rubles. There is, however, a 5% probability that the ship will not reach Amsterdam. It is possible to insure the goods for a premium of 800 rubles. Bernoulli calculated that Caius would only be prepared to pay the premium if his own capital did not exceed 5,043 rubles. If he had more capital, he would consider the premium too high. At a premium of 600 rubles, the critical level of capital is 20,478 rubles. It seems paradoxical that the poorer you are the more premium you are prepared to pay. However, the potential loss feels harsher if it takes up a large proportion of your capital.

Bernoulli was very satisfied with his own results and gave himself a slap on the back: "Though a person who is fairly judicious by natural instinct might have realized and spontaneously applied much of what I have here explained, hardly anyone believed it possible to define these problems with the precision we have employed in our examples. Since all our propositions harmonize perfectly with experience it would be wrong to neglect them as abstractions resting on precarious hypotheses."

It would clearly have come as no surprise to Bernoulli that his work was important for the economic sciences, but that his influence would extend to evolutionary biology and behavioral science certainly would have surprised him. He advised, for example, spreading goods that may be at risk over different consignments. This principle of "bet-hedging" or "not putting all your eggs in one basket" is often applied in evolutionary biology.

Bernoulli used mathematical methods to solve a wide variety of problems. That started with his analysis of flowing fluids, but was equally true of his development of the utility function. With hindsight, it is remarkable that Bernoulli became more famous among the public at large for a principle that was partly incorrectly used to explain the lift effect of wings than for his theory of utility, which plays such an important role in our daily lives.

References

Daniel Bernoulli, 1954. 'Exposition of a New Theory on the Measurement of Risk', *Econometrica* 22 (1), 23–36. English Translation of the original *Specimen Theoriae Novae de Mensura Sortis* from 1738.

Stephen C. Stearns, 2000. 'Daniel Bernoulli (1738): Evolution and Economics under Risk'. *Journal of Biosciences* 25 (3), 221–228.

Robert William Vivian, 2003. 'Solving Daniel Bernoulli's St Petersburg Paradox: The Paradox which is not and never was'. *South African Journal of Economic and Management Sciences* 6, 331–345.

Benjamin Franklin

Benjamin Franklin is mainly renowned for his classic experiment with a kite, with which he demonstrated that lightning is simply a matter of electricity. As well as being a scientist, Franklin was a printer, publisher, diplomat, and politician. As Deputy Postmaster General, he was responsible for mail traffic between the new world of the American colonies and the old world in Europe. In that capacity he had detailed maps drawn up of the warm Gulf Stream that flows from North America to Europe.

R. Schils, *How James Watt Invented the Copier: Forgotten Inventions of Our Great Scientists*, DOI 10.1007/978-1-4614-0860-4_5,
© Springer Science+Business Media, LLC 2012

Electricity

When he was a teenager, Benjamin Franklin started working as an apprentice at his brother James' printing and publishing house in Boston. But the two brothers did not get on and, in 1723, Benjamin packed up and headed for Philadelphia, where he found employment at various printing companies. His work was noticed by Governor William Keith, who promised him a contract to publish a new newspaper. Keith sent Franklin to London to buy the necessary machinery. Unfortunately, Keith's promise proved to be nothing more than hot air and Franklin found himself alone again.

Back in America, Franklin set up his own printing and publishing company. The company was a success and, within a short time, he had become one of Pennsylvania's most prominent figures. After about 20 years, he sold the business and could afford to dedicate himself entirely to science and politics, which he had occasionally dabbled in when he was younger. In 1747 he started to experiment with electricity and, 5 years later, published his book *Experiments and Observations on Electricity*.

As a scientist, Franklin is primarily associated with his famous experiment with the kite in a thunderstorm. The aim of the experiment was to show that lightning is simply an electrical discharge. Franklin claimed that, if the kite were struck by lightning, the electric discharge would travel down the line. At the bottom of the line there was a key, from which a spark would jump across to his hand. In all probability, the experiment only ever took place in Franklin's mind. Study of his records reveal considerable gaps in the information on when and where it took place. Furthermore, it has never proved possible to repeat the experiment with materials

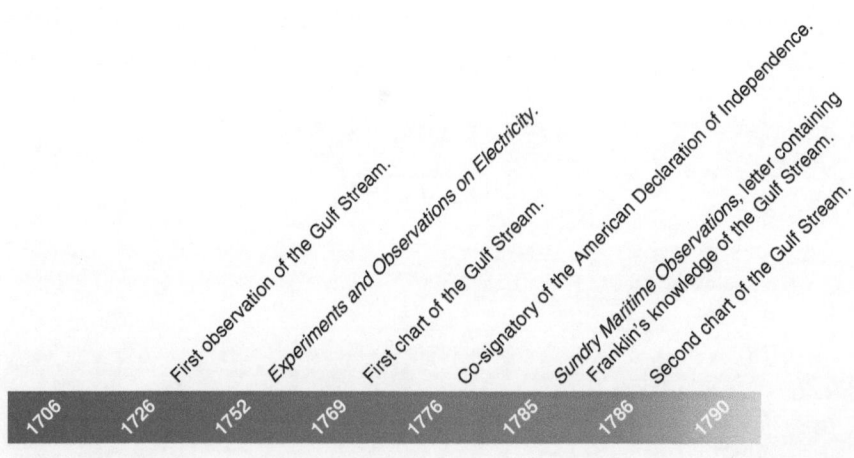

Benjamin Franklin

that Franklin had at his disposal in the eighteenth century. Nevertheless, many of his theories on electricity proved correct and he was later to design the first lightning conductor.

Franklin also made quite a name for himself as a politician and statesman. In 1757, he traveled to England to represent the state of Pennsylvania in a conflict with the Penn family over the colonial representation of the state. His stay was extended and, in the end, he represented four American states in England. His loyalty to England was severely tested after the overwhelming opposition in America to the Stamp Act, a tax imposed on the American colonies. Not long after, under great political pressure, Franklin had to leave England. Back in America, he devoted his efforts to the struggle for independence. He was elected to the Second Continental Congress and was a member of the Committee of Five that drew up the Declaration of Independence. He was also one of the signatories in 1776.

Gulf Stream

In 1726, after 2 years in London, the young Benjamin Franklin returned to Philadelphia. During the voyage, he noticed *Sargassum bacciferum*, a species of seaweed found in the tropics, floating everywhere in the water, from the west coast of England to the east coast of America. He also observed that the water had changed color and that the air had become warm and humid. As was clear from his journal, not everyone on board shared his observations: "The water is now very visibly changed to the eyes of all except the Captain and Mate, and they will by no means allow it; I suppose because they did not see it first." Although he was not aware of this himself, this was most likely Franklin's first encounter with the Gulf Stream.

More than 40 years after this first encounter, the Gulf Stream again demanded Franklin's serious attention. It was 1768 and he was back in London, now as Deputy Postmaster General. As the head of the American mail service, he was asked why it sometimes took the English mail packets up to 2 weeks longer to make the crossing from England to America than American merchant ships. At first, Franklin thought it must be a mistake or that he had misunderstood the question. After all, the American trade ships were often more heavily laden and had less well-trained crews. But he did consult his cousin Timothy Folger, captain of a whaling ship, who came from Nantucket Island. Folger did not have to think about the question for very long. He told Franklin that the time difference could very well be explained by the Gulf Stream. The whalers had discovered that their prey were usually either south of a certain line or north of another one, but rarely in between the two. The two lines marked the limits of the Gulf Stream. According to Folger, the American merchant vessels made use of this knowledge of the Gulf Stream and adjusted their courses accordingly, while it was an unknown phenomenon for the English.

During their voyages, the whalers regularly crossed the Gulf Stream. Folger told Franklin that they sometimes encountered English ships sailing to America right down the center of the stream. Folger did little to conceal his disdain for the arrogance of the English sailors: "We have informed them that they were stemming a current, that was against them to the value of 3 miles an hour; and advised them to cross it and get out of it; but they were too wise to be counseled by simple American fishermen."

Franklin realized immediately how important this was to shipping and had Folger draw the course of the Gulf Stream on a chart, which he then had printed and distributed. He also presented the chart to the English postal service, advising them to stay out of the Gulf Stream during the westward crossing. Unfortunately for the English seamen, the postal service did little more than take note of the information. Perhaps they did not trust Franklin because of his role in the confrontation between England and the American colonies. Others have alleged that Franklin himself withheld the chart because he did not want it to fall into the hands of the British navy.

But the chart alone was not enough for Franklin. He was so fascinated by the "river in the ocean" that, during the return voyage in 1775, he took the temperature of the water as often as 4 times a day. Because the temperature of the Gulf Stream is higher, the measurements helped to better determine its precise location. A year later, after the signing of the Declaration of Independence, Franklin traveled to France to seek support for the American Revolution. During the voyage, he once again tested the temperature and color of the water, and the presence of seaweed.

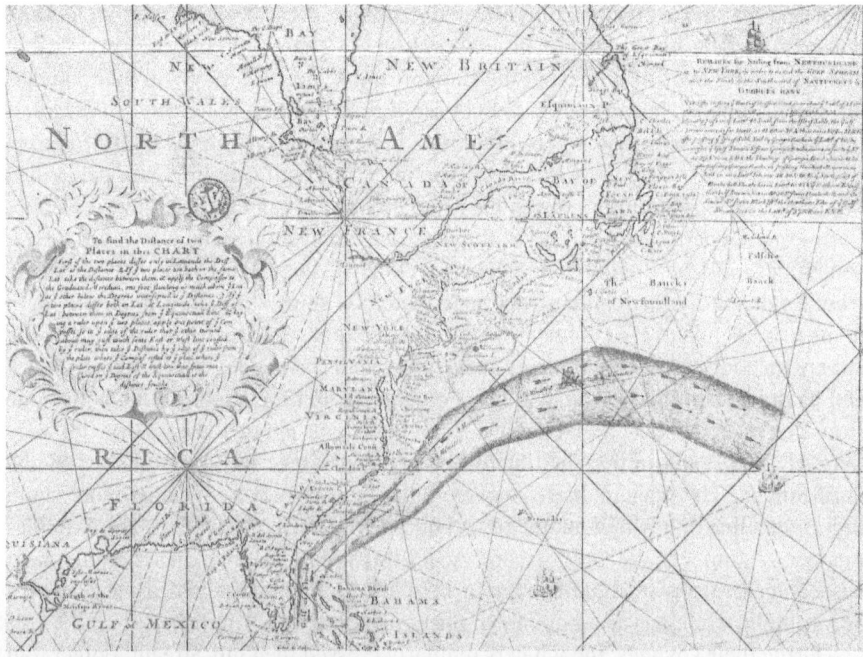

First edition of Franklin and Folgers' chart of the Gulf Stream from 1769. The chart also gives instructions on how to avoid the Gulf Stream

Second edition of Franklin and Folgers' chart of the Gulf Stream from 1786. This chart is not an exact copy of the first edition. The projection is different and the Gulf Stream itself is traced in a different way. For economic reasons, the publisher used the *top left* corner of the chart to show a map belonging to another article in the same publication. This led to misunderstandings, as the small map has nothing to do with Franklin's chart. It depicts John Gilpin's description of the annual migration pattern of herring. Another interesting feature of this chart is, in the *bottom right*, Neptune talking to Franklin

In 1785, Franklin wrote everything he had learned about the Gulf Stream in a letter to a French colleague. The letter, which contained what he called "Sundry Maritime Observations," is much more than just a description of the Gulf Stream, dealing with all kinds of other practical matters relating to shipping, including the "action of the wind" and the topic of sailing, as well as the various stages of sinking that a ship passes through after suffering a hole in its hull, and the various calamities that can be fatal to a ship during the crossing.

It cannot, of course, be claimed that Franklin discovered the Gulf Stream. Cartographer William Gerard De Brahm also studied the phenomenon. Between 1772 and 1776, he published charts of the Atlantic Ocean. De Brahm had been commissioned to map Florida, but did not restrict himself to the terrestrial aspects and also charted the coastal waters. And long before Franklin or De Brahm, the Vikings referred to strange currents close to the coast of North America. The first direct reference to the Gulf Stream, however, dates from 1513, when the Jesuit Ponce de Leon described what he calls the "Florida current," the part of the Gulf Stream that flows along the coast of Florida: "A current such that, although they had great wind, they could not proceed forward, but backward and it seems that they were proceeding well; at the end it was known that the current was more powerful than the wind." The first sketch of the Gulf Stream, by Althanasius Kircher, dates from 1665, but was of no use in navigation. Franklin's, or rather Folger's, chart was the first accurate representation that could be applied in practice. The chart described how to make use of the current when sailing eastwards and how to avoid it when moving to the west.

Today, interest in the Gulf Stream is much more focused on climate change than on navigation. The warm current is part of the thermohaline circulation, which transports warm water from the equator to the Arctic seas. As it cools off, the water sinks to a great depth, after which it flows back southwards. The Gulf Stream ensures that the climate in Western Europe is milder than elsewhere at the same latitude. Some climate models predict that the Gulf Stream will slow down, or even come to a complete halt, but it is extremely uncertain whether this will happen and when. If the circulation were to stop completely, temperatures in Northwest Europe would fall by between 2° and 4°. In this respect, too, Franklin was way ahead of his time: he is alleged to have proposed altering the course of the Gulf Stream to freeze the English, his country's opponents in the American Revolution.

References

Benjamin Franklin, 1786. 'A Letter from Dr. Benjamin Franklin, to Mr. Alphonsus le Roy, Member of Several Academies at Paris. Containing Sundry Maritime Observations. At Sea, on board the London Packet', Capt. Truxton, August 1785.
Franklin Bache, 1936. 'Where is Franklin's first chart of the Gulf Stream?' *Proceedings of the American Philosophical Society* 76 (6), 731–741.
Louis de Vorsey, 1976. 'Pioneer Charting of the Gulf Stream: The Contributions of Benjamin Franklin and William Gerard De Brahm'. *Imago Mundi* 28, 105–120.
Philip L. Richardson,1980. 'Benjamin Franklin and Timothy Folger's First Printed Chart of the Gulf Stream'. *Science* 207 (4431), 643–645.

Joseph Priestley

Scottish chemist Joseph Priestley is credited with the discovery of ten new gases including, as the jewel in his crown, oxygen. Priestley's early efforts led him to devise a method of adding carbon dioxide to water. This enabled him to imitate natural mineral water, laying the foundation for a new industry.

R. Schils, *How James Watt Invented the Copier: Forgotten Inventions of Our Great Scientists*, DOI 10.1007/978-1-4614-0860-4_6,
© Springer Science+Business Media, LLC 2012

Oxygen

Joseph Priestley was born and bred in a Calvinist community. His main interests therefore lay more in the field of theology than in the natural sciences. Throughout his adult life he was active as a clergyman in various local parishes. He was also a very passionate participant in controversial debates on religion and politics in England and elsewhere. Priestley's interest in science was given a considerable boost by his meetings and correspondence with Benjamin Franklin.

In 1766, Priestley was admitted to the Royal Society and a year later he published his first great scientific work, *The History and Present State of Electricity, with Original Experiments*. In the years that followed, he conducted intensive research into gases. Between 1774 and 1786, he published *Experiments and Observations on Different Kinds of Air* in six volumes. Until then, gases – or "airs," as they were called at the time – were divided into roughly three categories: normal air, "fixed" air (carbon dioxide), and flammable air (hydrogen). Priestley discovered ten new gases, including various nitrogen oxides, ammonia, sulfur oxide, nitrogen and, of course, oxygen. The latter is generally seen as his most important discovery. In 1774, Priestley discovered that heating the mineral mercuric oxide releases a gas that causes a candle to burn brighter and a mouse to live longer. He called the gas "dephlogisticated air," in accordance with the phlogiston theory devised in the seventeenth century by the German Georg Ernst Stahl. The theory states that combustible materials contain a substance known as phlogiston that is transformed into fire when heated.

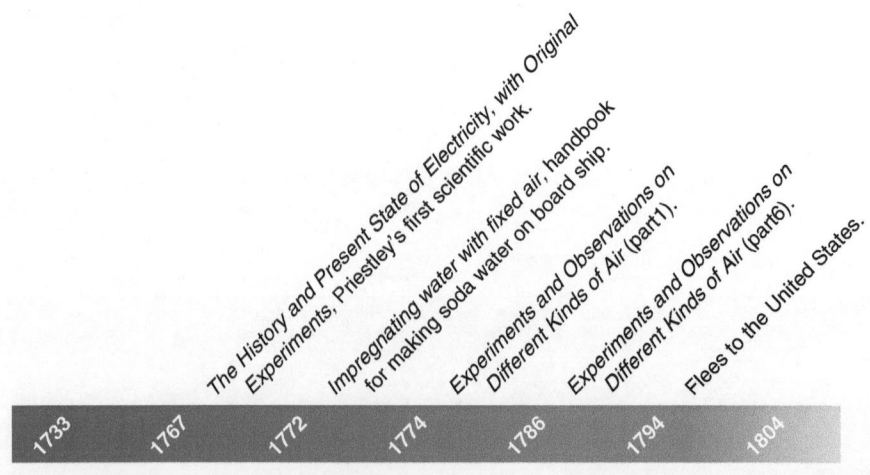

Joseph Priestley

Priestley also described the effect of inhaling oxygen on himself, a privilege that he said he had previously only granted to two mice. "The feeling of it in my lungs was not sensibly different from common air, but I fancied that my breast felt peculiarly light for some time afterward. Who can tell but that, in time, this pure air may become a fashionable article of luxury."

After his discovery, Priestley traveled to the mainland of Europe, where he came into contact with Antoine-Laurent Lavoisier. In the years that followed, Lavoisier conducted similar experiments and introduced the name oxygen. While Lavoisier further elaborated the active nature of oxygen during combustion and respiration, Priestley continued to adhere to the obsolete phlogiston theory.

Although Priestley and Lavoisier are attributed with the discovery of oxygen, Swedish chemist Carl Wilhelm Scheele had already established in 1771 that normal air comprises a quarter "fire air" (oxygen) and three-quarters "vitiated air" (nitrogen). And a 100 years earlier, John Mayow described how only a part of the air was needed for us to live.

Priestley's scientific activities did not stop him from remaining active in religion and politics. But his open support for the American Revolution made him unpopular and on July 14, 1791, a horde of opponents destroyed his home and laboratory. Priestley and his family took flight, first to a safe place in England and later, in 1794, for good to the USA, where he was received as a celebrity.

Soda Water

In 1767, Priestley moved to Leeds and found himself living next door to a large brewery. It provided him with a convenient source of carbon dioxide. "…living for the first year, in a house that was contiguous to a large common brewery, so good an opportunity produced in me an inclination to make some experiments on the fixed air that was constantly produced in it. Had it been not for this circumstance, I should probably never have attended to the subject of air at all."

He began with a few exploratory experiments, holding a burning candle or glowing woodchips above the fermentation vat. He concluded that all combustible materials are extinguished in the layer of air 20–25 cm above the fermenting liquor. Then, one evening, he placed a shallow dish filled with a little water above the fermenting beer. The next morning, he tasted the water and noted that it had a pleasant, sharp taste leading him to conclude that it contained carbon dioxide. He achieved the same effect more quickly by pouring water between two glasses that he held in the carbon dioxide layer. He described the result as exceptionally pleasant sparkling water, hardly distinguishable from genuine Pyrmont mineral water. An accident involving spilled beer brought his experiments in the brewery to an abrupt end. Though Priestley knew that there were other ways to produce carbon dioxide, by burning charcoal, heating lime, or applying acid to lime or marl, he saw little reason at that moment to explore ways of imitating natural mineral water any further.

In eighteenth-century England, social life outside London centered around spa towns like Bath and Tunbridge Wells. The wealthier members of English society would gather in these places, where a culture developed of drinking and bathing in the

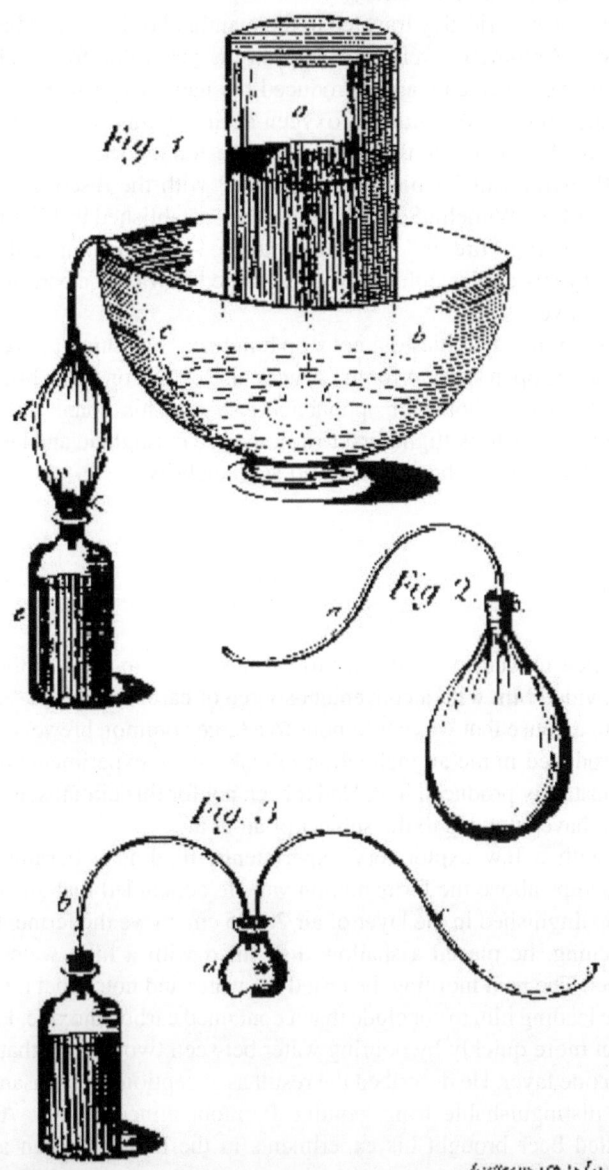

Priestley's apparatus that enabled mineral water to be produced on board ship. The generator bottle contains lime and sulfuric acid. The carbon dioxide is introduced into the water using a bladder and tube

Schweppes mineral water from the early nineteenth century. The bottle bears the name of the street where the London factory was located (Margaret Street)

mineral water. In between taking the waters, you could relax in the numerous coffee houses, shops, theaters, and libraries. On the European continent, too, similar centers emerged, such as Spa in Belgium and Pyrmont in Germany. Even more than in England, health was the main priority and the benefits of mineral water were widely proclaimed. Pyrmont water, for example, was alleged to help against a whole list of ailments, including a weak stomach, indigestion, nervous complaints, and heart problems. Yet, healthy as it was, staying at a spa was an expensive business. To allow the less well off also to benefit from the healing properties of mineral water, it was bottled and sold to the man in the street.

Scientists focused their attention on the gases and minerals in the water. In *An Experimental Enquiry into the Mineral Elastic Spirit of Air, contained in Spa Water*, William Brownrigg described how he had tied an empty bladder around the neck of a bottle of spa water and heated the water to around 40°C. In this way, he collected a quarter of a liter of the "fixed air," carbon dioxide. Research showed that mice did indeed not survive in the gas extracted from the spa water, leading many to conclude that carbon dioxide was the main element in spa water affecting the health.

Four years later, Priestley's interest in carbonated water was aroused once again during a dinner with the Duke of Northumberland. Another of the table guests was Charles Irving, a British ship's doctor concerned with the problem of how to preserve the quality of water during long sea voyages. Irving thought the solution lay in distilled water, but Priestley claimed that this lacked the freshness of mineral water. He proposed improving the taste by adding carbon dioxide. Priestley also suspected that carbonated water could prevent scurvy. His proposals were submitted to the Admiralty and, soon afterwards, two ships were fitted with equipment to produce carbonated water. One of the ships was the *Discovery*, in which James Cook sailed on his second voyage to the South Pacific. Cook, however, was so successful at finding fresh drinking water that he hardly needed to use Priestley's equipment, if at all.

Priestley published a detailed description of his method in a special document for the Admiralty, *Impregnating Water with Fixed Air; In order to communicate to it the peculiar Spirit and Virtues of Pyrmont water, And other Mineral Waters of a similar Nature*. The equipment consisted of a generator bottle containing sulfuric acid and lime. The bottle was connected to an absorption vessel with a flexible leather tube. To simulate the taste of Pyrmont mineral water, a few drops of *Tinctura Martis cum spiritu salis* – ferric chloride in a solution of hydrochloric acid – were added. A bladder was placed between the generator and the absorption tank to regulate the flow of carbon dioxide. Priestley claimed the bladder was necessary to allow the pressure to be increased. Others were afraid, however, that using a bladder would give the water the taste of urine, and devised mechanical pumps to introduce the carbon dioxide into the water under pressure. As an aside, Priestley noted that his apparatus could also be used to inject new life into beer that had gone dead.

As interest in carbonated water increased, Priestley's apparatus was soon joined by a wide range of others. Most of them, however, were designed only for use in households or dispensing chemists. It was not until 1781 that carbonated water began to be produced on a large scale, with the establishment of companies specialized in producing artificial mineral water. The first of these factories was built in Manchester, England, by Thomas Henry. To increase the pressure, he replaced the bladder in Priestley's system with large bellows. Backed up by mineral water's healthy image, dozens of successful companies were set up, including that of Jacob Schweppe, which still exists today. Schweppe started producing artificial seltzer water in Geneva, and soon had factories in London and elsewhere in England.

The name "soda water" was introduced after chemist Richard Bewley suggested adding a little soda (sodium carbonate). In the early nineteenth century, bottled water became increasingly popular and was no longer drunk pure or solely for health reasons. Gradually, other tastes were added, including lemon, ginger, and quinine (tonic) and, in 1886, John Pemberton developed Coca-Cola.

Although Priestley is "the father of the soft drink," he never benefited financially from his invention. He did, however, receive scientific recognition in 1773 when he was awarded the Copley Medal by the Royal Society. Remarkably, the prize was awarded on the basis of the misconception that carbonated water could prevent and cure scurvy.

References

Joseph Priestley, 1772. *Impregnating Water with Fixed Air; In order to communicate to it the peculiar Spirit and Virtues of Pyrmont water, And other Mineral Waters of a similar nature.* Johnson, 22 pp.

William Back *et al.*, 1963. 'Bottled Water, Spas, and Early Years of Water Chemistry'. *Journal of the National Water Well Association* 33 (4), 605–614.

W.A. Campbell, 1983. 'Joseph Priestley's soda water'. *Endeavor, New Series* 7 (3), 141–143.

James Watt

James Watt is best known as the inventor of the steam engine and driver of the Industrial Revolution. This reputation is not entirely deserved, as his invention was actually an improvement on a steam engine invented half a century earlier. Watt was, however, the real inventor of the copying machine.

R. Schils, *How James Watt Invented the Copier: Forgotten Inventions of Our Great Scientists*, DOI 10.1007/978-1-4614-0860-4_7, © Springer Science+Business Media, LLC 2012

Steam Engine

James Watt learned the basic principles of instrument-making in his father's workshop. After studying in Glasgow and London, he opened his own workshop at the University of Glasgow in 1757. There he made a variety of instruments, especially compasses and balances. In the years that followed, Watt began to experiment with steam, but was unable to make a working model. Around that time, he discovered that the university had a model of a Newcomen steam engine, which was in London waiting to be repaired.

The Newcomen engine did not work on steam pressure, but on the vacuum created by condensing steam. Watt had the model brought to Glasgow, where he got it running again. Much more importantly, he made an essential improvement that radically reduced the loss of energy. In brief, he added a separate chamber in which the steam condensed, retaining the temperature of the cylinder. In 1769, Watt applied for a patent on "A new invented Method of Lessening the Consumption of Steam and Fuel in Fire Engines."

At first, Watt had difficulty in getting someone to produce his new engine. In 1775, however, his luck changed when Michael Boulton acquired the patent. A year later, the first two "Boulton & Watt" steam engines were in operation, one as a water pump in a coal mine and the other as an air pump for the furnaces of an iron foundry.

In the years that followed, Watt and his colleagues improved the engine on a number of essential points. Fitting a crankshaft, for example, enabled them to transfer the reciprocating motion of the piston into rotational movement. The engine could now be used to process grain or cotton. Watt also modified the design so that the piston was driven alternately by pressure from both sides.

By 1800, Boulton & Watt had supplied some 500 machines. Despite all of Watt's modifications, the efficiency of these engines was never much higher than 2%.

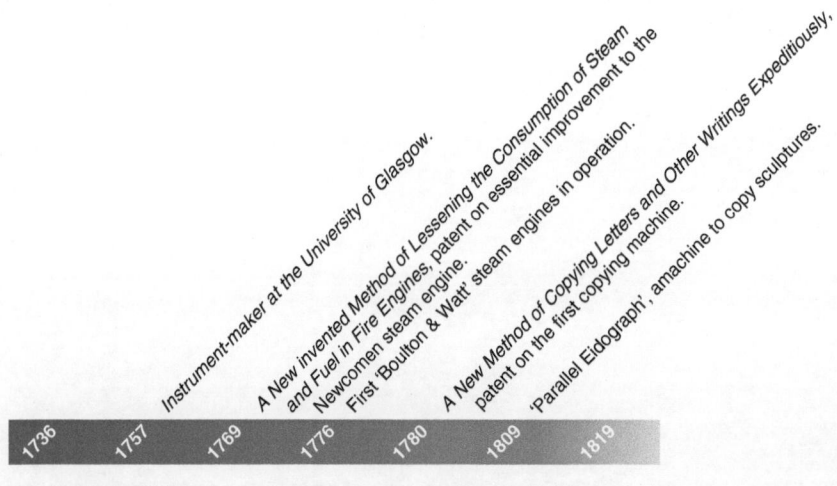

James Watt

In the course of the nineteenth century, this was gradually increased to around 17% by, for example, using steam under high pressure.

To express the power of his steam engines, Watt invented the term "horsepower," based on how many horses his clients would save if they purchased one. For many years, horsepower was the international standard unit for measuring power, until it was replaced in 1960 – by the Watt.

Copying Machine

In Redruth, Cornwall, James Watt sat gloomily looking at piles of paperwork. Together with his partner, Michael Boulton, Watt had a flourishing company that supplied steam-driven water pumps to the local mines, where the miners literally worked up to their ankles in water. The orders flooded in. But the downside of their success was the massive volume of paperwork. Letters, detailed construction drawings and bills lay around waiting to be copied. At that time, copying documents was no simple matter. Companies employed clerks to write out the most important documents word for word. Finding suitable clerks was one of Watt's greatest problems. The mistakes they made copying the documents by hand drove

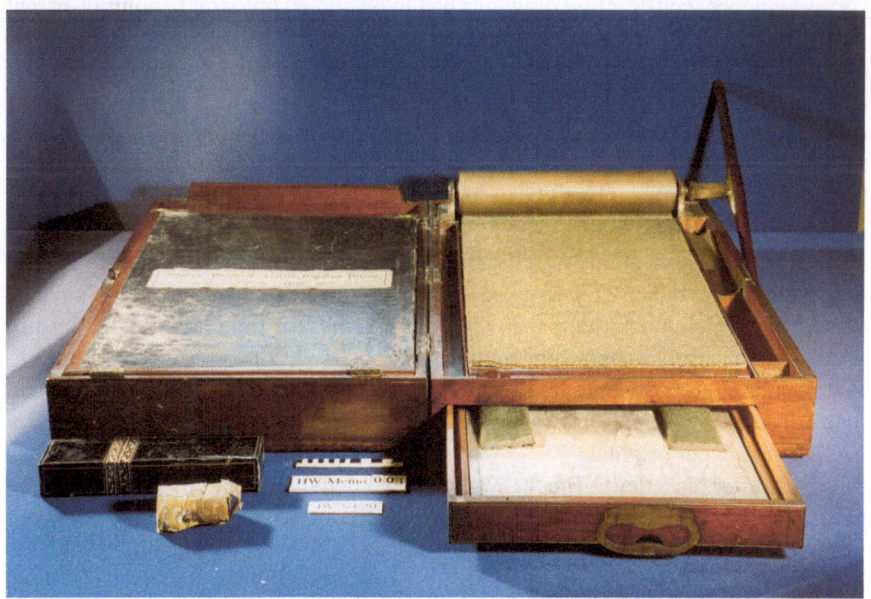

A portable and collapsible copying machine by James Watt

him to distraction. So he invented a machine that could copy letters and other documents more quickly and accurately.

In 1779, Watt shared his secret with his friend Joseph Black, a Scottish chemist: "I have lately discovered a method of copying writing instantaneously, provided it has been written within twenty-four hours…. It enables me to copy all my business letters." Watt immediately realized the commercial value of his invention, and Black was one of the few to whom he entrusted his secret. Watt and Black were both members of the Lunar Society, a group of prominent scholars in the Birmingham area who met every Monday around Full Moon between 1765 and 1813. Although it was an informal association, the Lunar Society was second only to the Royal Society when it came to scientific influence. Besides Watt and Black, its members included Joseph Priestley, Matthew Boulton, Erasmus Darwin, and Benjamin Franklin, the latter of course corresponding mainly by letter. Their shared goal was to apply science to production, transport, and other social activities. The Lunar Society network was used to bring Watt into contact with a number of prominent figures who would certainly be interested in a copying machine. The widely varying group who did show an interest included economist Adam Smith, author of the classic *The Wealth of Nations*, banker William Forbes, and physician William Cullen, all from Edinburgh.

Watt's invention was based on a relatively simple principle. The original had to be written using a kind of gelatinous ink. It was pressed against the paper on which it was to be copied, which had been slightly moistened, and placed in a press or passed between two rollers. This pressed the ink of the original through the copy paper, rendering the text visible on the other side. Although this was acceptable for correspondence, construction drawings had to be copied onto thick, nontransparent paper. The drawings were then stamped as "REVERSE."

The members of the Lunar Society also helped Watt to solve a number of technical problems. Developing suitable ink, for example, was a time-consuming chemical puzzle. The ink used for the original document had to be thick enough, without smudging. When it came into contact with the moist copying paper, a part of the ink had to be liquid so that it would pass through the paper under pressure, but without running at the edges.

In his patent application, Watt lists the ingredients of the ink, which include mineral water, gum arabic, Aleppo galls, and green vitriol (iron sulfate). Watt's recipe contained more gum and galls than the normal ink of the time. Yet the quality still proved less than optimal and the copies rather pale. On the advice of James Keir, Watt increased the gum and galls even further, which proved to be an improvement on all fronts. The quality of both original and copy was better, and more copies could be made per original.

A year after his letter to Black, Watt was awarded the patent on "A New Method of Copying Letters and Other Writings Expeditiously," after which his copying machine could go into production. The copier was a great success and Watt sold 200 in the first year. The invention soon became popular beyond the borders of Scotland and England. Thomas Jefferson, coauthor of the American Declaration of Independence and third president of the USA, was more aware than most of the

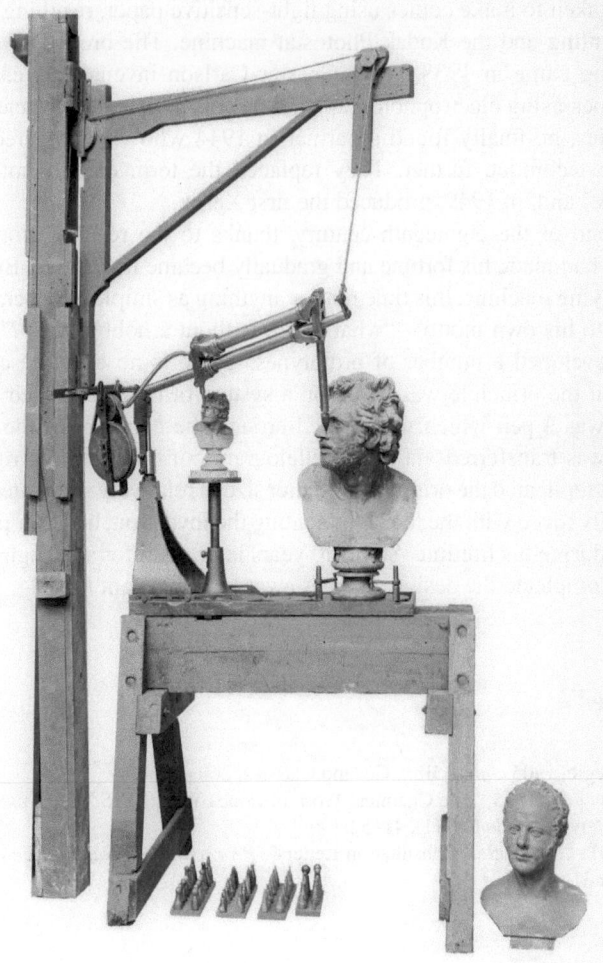

Benjamin Cheverton's machine to copy sculptures, based on a design by James Watt

value of keeping careful records of public documents. He used several versions of Watt's copying machine to do so.

The copying machines were produced by James Watt & Company, in which Keir and Boulton were partners. In the nineteenth century, the machines gradually became a normal office fixture. Alongside standing copiers, James Watt & Company later produced a portable version. In the course of time, more and more versions appeared on the market, some produced by competitors, but for many years it was still only possible to copy "freshly written" documents.

With the invention of carbon paper, especially in combination with the typewriter, Watt's invention gradually receded into the background. Later, the first attempts were undertaken to make copies using light-sensitive paper, resulting in techniques like blueprinting and the Kodak Photostat machine. The breakthrough to modern photocopying came in 1938, when Chester Carlson invented an easy method of making copies using electrophotography. After his idea had been rejected by some 20 companies, he finally found a partner in 1944 who was prepared to help him develop the technique further. They replaced the term electrophotography with "xerography" and, in 1949, produced the first Xerox.

By the end of the eighteenth century, thanks to the revenue from his patents, James Watt had made his fortune and gradually became less active. But he invented one last copying machine, this time not for anything as simple as paper, but for sculptures. True to his own motto – "what is life without a hobby-horse?" – in his final years, he developed a number of prototypes. There is no exact description of the machine, but the principle was based on a system of parallel hinged arms. On one side, there was a pen which was guided around the contours of the original. The movement was transferred via the parallelogram construction to a rotating cutting element that replicated the original in smaller size in relatively soft material. Although Watt certainly toyed with the idea of patenting the invention, he never perfected it for production during his lifetime. Some 20 years later, sculptor and engineer Benjamin Cheverton completed the design and was awarded the patent in 1844.

References

Andrew Carnegie, 1905. *James Watt.* Cosimo Classics, 260 pp.
Jennifer Pugh *et al.,* 1985. 'The Chemical Work of James Watt, F.R.S.' *Notes and Records of the Royal Society of London* 40 (1), 41–52.
J. Dallas, 2001. 'The Cullen Consultation Letters'. *Proceedings Royal College of Physicians of Edinburgh* 31, 66–68.

Edward Jenner

The worldwide eradication of small-pox is one of the greatest milestones in modern medicine. More than 200 years ago, Edward Jenner took the first important step when he developed the smallpox vaccine. Yet, long before he developed the vaccine, Jenner acquired scientific fame for his research into the exceptional nesting habits of the cuckoo.

R. Schils, *How James Watt Invented the Copier: Forgotten Inventions of Our Great Scientists*, DOI 10.1007/978-1-4614-0860-4_8,
© Springer Science+Business Media, LLC 2012

Smallpox Vaccine

Edward Jenner was a real country boy, born and bred in Berkeley, in the English county of Gloucestershire. At the age of 14, he was apprenticed to a local doctor's practice, where he laid the basis for his medical training. Seven years later, he left to complete his studies at St. George's Hospital in London, where he was taught by John Hunter. In London, Jenner stayed in Hunter's house, where his education continued unabated. In the autopsy room in the basement of Hunter's house, he learned everything about anatomy and physiology, 6 days a week, starting at 6 o'clock in the morning. Despite Hunter's earnest pleas for him to stay in London, Jenner left the big city as soon as possible after finishing his studies. Back in Berkeley, he set up his own practice.

In the eighteenth century, smallpox was very common and a major cause of death, especially among children. Berkeley, too, had its share of victims and, in 1788, Jenner found himself fighting a smallpox epidemic. He noted that patients who worked with cows never contracted real smallpox, but only the much milder cow pox. Jenner developed a hypothesis that people who had been infected with cow pox would never contract smallpox. He thought that he could protect people for catching smallpox by deliberately giving them cow pox. Jenner got his opportunity 8 years later, when he persuaded his gardener to allow him to use his son as a guinea pig. First, he extracted a little fluid from the blisters of a patient suffering from cow pox, which he then injected into two small cuts on the healthy guinea pig's left arm. The boy developed cow pox, but it did not make him overly ill. Six weeks later,

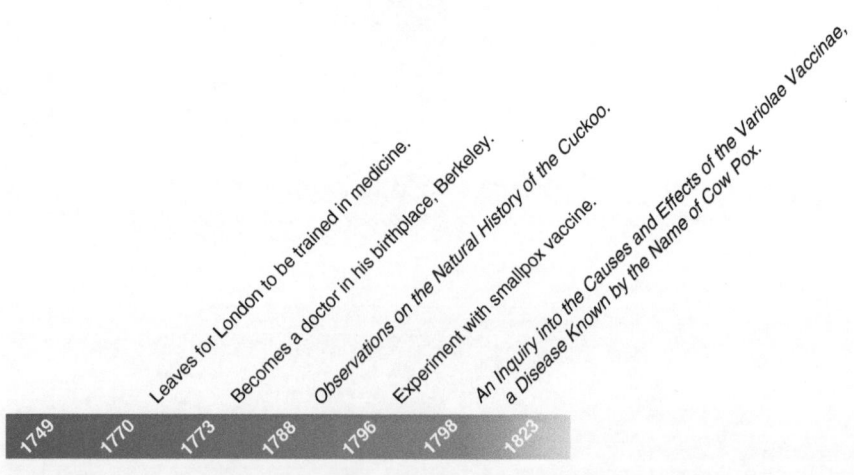

Edward Jenner

Jenner injected the boy with the smallpox and, to everyone's great relief, he did not become ill, as Jenner had predicted.

He described his theory and the results of his experiment in a paper for the Royal Society, but its eminent members were not so easily persuaded. Moreover, the criticism came not only from the scientific community: the church found it a heathen practice to infect a human with animal tissue. Cartoons even appeared depicting vaccinated people with cows' heads sprouting out of their hands. But Jenner resolutely continued his experiments, vaccinating more children, including his own 11-month-old son. Eventually, the Royal Society accepted his proof and he published *An Inquiry into the Causes and Effects of the Variolae Vaccinae, a Disease Known by the Name of Cow Pox*. His method spread like wildfire, and what started with country-lore which said that milkmaids who caught cow pox could not catch smallpox, ended over 200 years later with the worldwide eradication of the smallpox virus.

Cuckoo Chicks

On June 19, 1787, Edward Jenner saw with his own eyes how a 1-day-old cuckoo chick (*Cuculus canorus*) unceremoniously shoved a hedge sparrow (*Prunella modularis*) out of its own nest. The mystery had finally been solved. The parasitical nature of the cuckoo's nesting behavior had been known for centuries, but how the young of the host mother disappeared from the nest had always been a riddle. Aristotle wrote that the cuckoo only forced the other chicks out of the nest once they were full-grown. Some researchers in Jenner's time thought that the mother cuckoo ate the legal residents of the nest or that they simply starved because the cuckoo chick devoured all the food. An even more gruesome suggestion was that the host mother herself killed her own chicks and then allowed the cuckoo chick to devour them, because she found the cuckoo more attractive than her own brood.

Jenner had been fascinated by nature since his childhood, and he loved the countryside with his heart and soul. He would search for eggs in woods and meadows, and scoured the estuary of the River Severn for fossils. He had been forced to leave his beloved surroundings temporarily for London to complete his medical training, but as soon as he had been awarded his diploma, he quickly returned to his birthplace. Besides the homesickness that drove him home, Jenner had no desire at all to develop himself further in London. However, his departure by no means meant that he lost contact with his tutor, John Hunter. The man who had been responsible for Jenner's training did not lose sight of his pupil. They exchanged letters regularly until Hunter's death. Jenner's letters have unfortunately been lost, but it is clear from Hunter's letters that Jenner's education in the natural sciences essentially continued.

In some respects, Jenner became an extension of Hunter, a remote laboratory. Hunter instructed Jenner to collect everything that had anything to do with animals and send it to London. Bats, crows' nests, swallows, herrons, hedgehogs, and lizards, whole or in parts, everything found its way to Hunter. Jenner also conducted experiments for Hunter and sent the results to London. For his part, Hunter made sure that Jenner had the right measuring equipment to make his observations accurately. In the meantime, Jenner continued with his own practice. In return for Jenner's efforts, Hunter would sometimes give him advice on how to treat his patients.

In addition, Jenner also conducted his own research into the behavior of birds, especially cuckoos. But it was Hunter again who proposed that he record his ethological study of the cuckoo in writing so that he would be admitted to the prestigious Royal Society. It is even doubtful whether Jenner would have finished the job at all without the encouragement of his former tutor. Jenner observed the nests of different species of bird in which the cuckoo laid her eggs, devoting special attention to the hedge sparrow. Every time, he discovered a veritable graveyard of dead chicks and eggs, broken or whole, on the ground beneath the nests, while in the nest itself a single cuckoo chick would be stuffing itself with all the food brought back by the mother. In the first instance, he concluded that it was almost certainly the mother herself who expelled the eggs from the nest, as a young cuckoo could never have sufficient strength. But he soon had to change his mind, after seeing with his own eyes how a cuckoo chick committed the murderous act itself.

The small cuckoo first uses its wings and body to get the egg or the chick on to its back. It then works its way backwards until it reaches the side of the nest where, with a final effort, it pushes the victim over the edge. Jenner also concluded that, lacking well-developed eyesight, the cuckoo uses its wingtips to feel around the nest for eggs or other chicks.

The cuckoo

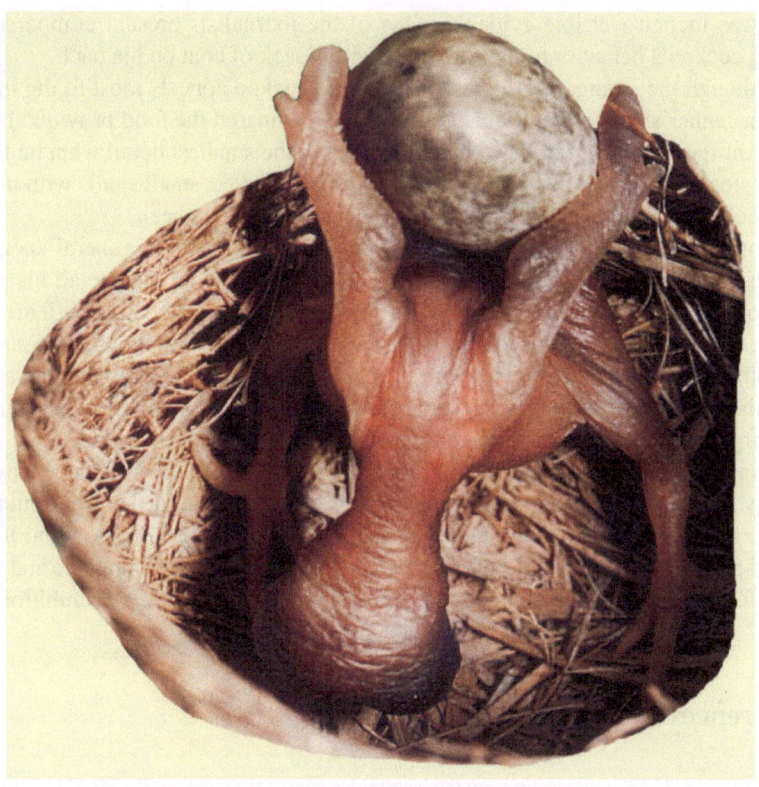

A cuckoo chick in action

With Hunter's motto – "Why think? Why not try the experiment?" – in the back of his mind, Jenner tried to confirm his findings. If he found unbroken eggs, he would replace them in the nest, only to find them on the ground again the following day. He succeeded in immobilizing the cuckoo chick, using small lead weights, as a result of which the original chicks and eggs remained unharmed. He also placed heavier birds in the nest, which the cuckoo was unable to push out. Whenever he forced the cuckoo to continue to share the nest with others, it was continually in a state of unrest.

Through detailed observations, Jenner discovered that the cuckoo chick has a kind of depression in its back, which appears to be a modification to enable it to carry an egg more easily on its back. After about 12 days, the depression disappears and the back takes on the same form of that of other birds.

After publication of Jenner's *Observations on the Natural History of the Cuckoo* in 1788, by no means everyone was convinced that he was right. He was inundated with criticisms of his claim that that such a small bird would be capable of such a violent deed. In scientific journals, experts did little to disguise their disdain. But in 1921, ornithologist Edgar Chance presented the first photograph of a cuckoo in action.

This was incontrovertible evidence. One of the journalists present compared the young cuckoo's behavior to that of a man with a sack of coal on his back.

Although the strange nesting behavior of the cuckoo appeals most to the imagination, Jenner's studies were much broader. He compared the food provided by the different species of stepmothers, and described in the smallest detail what he found in the stomachs of cuckoo chicks: various flies and beetles, small snails with unbroken shells, grasshoppers, caterpillars, and a piece of a broad bean.

Jenner later conducted research into how migratory birds in general spent the winter, but had less and less time to spend on that as he had to devote all his attention to smallpox. Jenner's study of migratory birds was not published until after his death, by his nephew, who had also been involved in many of the observations.

Although Jenner's research into the cuckoo had little to do with his work on smallpox, there is a great similarity in the method of working: both are perfect examples of sharp observation and experimentation, of precision and perseverance.

Hunter died while Jenner was still studying smallpox, but his influence was still clearly noticeable. Some authors have gone as far as to suggest that, if Hunter and Jenner had not remained in contact after the latter's departure from London, Jenner would probably not have come any further than running a doctor's practice and making a few notes on ornithology and anatomy; simply through a lack of ambition.

References

Edward Jenner, 1788. 'Observations on the Natural History of the Cuckoo. By Mr. Edward Jenner in a Letter to John Hunter, Esq. F.R.S.' *Philosophical Transactions of the Royal Society of London*, Vol. 78, 219–237.

Lloyd Alan Wells, 1974. '"Why Not Try the Experiment?" The Scientific Education of Edward Jenner'. *Proceedings of the American Philosophical Society*, Vol. 118, No. 2, 135–145.

E.L. Scott, 1974. 'Edward Jenner, F.R.S., and the Cuckoo'. *Notes and Records of the Royal Society of London*, Vol. 28, No. 2, 235–240.

David Bardell, 1996. 'Nestling cuckoos to vaccination: A commemoration of Edward Jenner'. *Biology in History*, Vol. 26, No. 11, 866–871.

John Dalton

John Dalton's reputation is largely
based on his atomic theory. However,
he made his scientific debut with an
analysis of color blindness, which –
as he had discovered shortly before –
he himself suffered from.

R. Schils, *How James Watt Invented the Copier: Forgotten Inventions
of Our Great Scientists*, DOI 10.1007/978-1-4614-0860-4_9,
© Springer Science+Business Media, LLC 2012

Atoms

John Dalton was born into a Quaker family in the Lake District, in England. He soon attracted attention because of his enormous eagerness to learn and his enthusiasm for natural scientific phenomena. At the age of 12, he was already teaching, first at his own school and, a few years later, at a larger school in nearby Kendal. At the same time, he was being tutored in classical languages and mathematics by two local scholars, Elihu Robinson and John Gough. Both were amateur meteorologists and Dalton was soon infected with their enthusiasm for everything to do with the weather. In 1787, he started keeping a meteorological journal, which he kept up until he died. The journal eventually contained some 200,000 observations.

In 1793, Dalton moved to Manchester where he started teaching mathematics at New College, a dissident institution that accepted nonconformists who had been refused admittance to Oxford or Cambridge. There, Dalton wrote on an extremely wide variety of subjects, which were often in one way or another connected to his meteorological observations, including rain and dew, heat, the color of the air, steam, and the reflection of light. It was here that he first developed his interest in gases, which would eventually lead to his atomic theory. Dalton studied the capacity of air to absorb water vapor, and the relation between air pressure and temperature. He discovered that the total pressure in a gas mixture is equal to the sum of the partial pressures of the individual gases in the mixture. This was later known as Dalton's law.

Although there are varying accounts of how Dalton continued his research, it is very probable that he eagerly sought evidence to support his gas law, which was the subject of fierce criticism. The long period between his first lecture on relative atomic weights in 1803 and the eventual publication of *A New System of Chemical Philosophy* in 1808 obscured the situation even further.

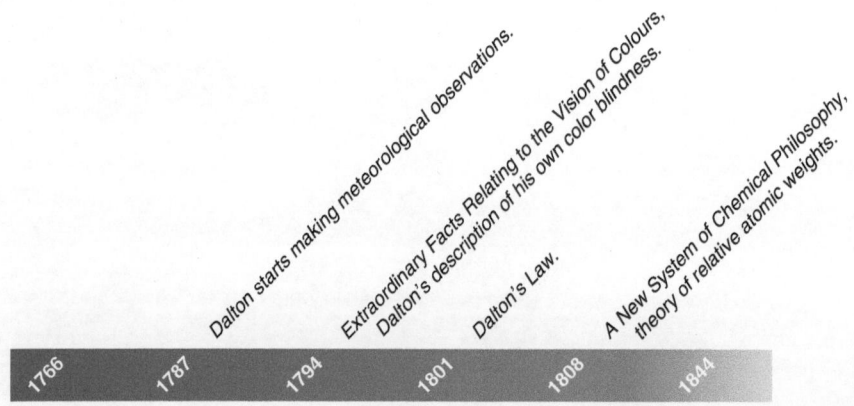

John Dalton

Dalton assumed that, in a gas mixture, identical atoms repel each other, while dissimilar atoms have no reciprocal effect. This later proved to be incorrect, but it did help him to abandon the classical notion that the atoms of all substances are identical. Dalton calculated the relative atomic weights of hydrogen, oxygen, and carbon by measuring the ratio in which they combined. He discovered that they react to each other in small, fixed ratios of 1 to 1, 1 to 2, or 2 to 1. Dalton's calculations proved correct in many cases, but not always. For example, he formulated methane as CH_2 instead of CH_4, and water as HO rather than H_2O. Despite these inaccuracies, Dalton's atomic theory allowed chemistry to take an enormous step forward. He also laid the basis for chemical notation with a graphic presentation of 21 elements.

Color Blindness

It was the geranium (*Pelargonium zonale*) in his school garden that made John Dalton realize that he saw colors differently to most people. To them, the flower was pink, but to Dalton it was sky-blue in daylight and almost yellow with a tint of red in candlelight. He observed that what other people called red was to him "little more than a shade or defect of light." Where they distinguished between the colors orange, yellow, and green, he saw only different shades of yellow. Dalton was color blind, a disorder that had never before been properly described.

Dalton presented his findings in his first scientific paper for the Manchester Literary and Philosophical Society in 1794, under the title *Extraordinary Facts Relating to the Vision of Colours*. He had been admitted to this respected club after becoming a teacher of mathematics and physics at the Presbyterian New College in Manchester. In the paper, he described not only his own color blindness, but that of others. He also presented a hypothesis of what caused it: he believed that the liquid medium in his eyeball was colored slightly blue, so that he could not see the longer wavelengths of red colors.

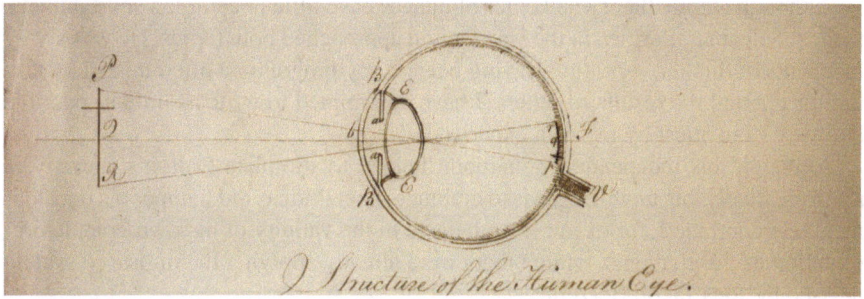

The structure of the human eye, according to Dalton

Among his closest associates, only his brother proved to have the same defect. Further investigation showed that it was not entirely unique: in a class of 25 children, two proved to see colors in the same way as Dalton. He never heard of any women with similar defects, which correlates surprisingly well with what we now know about the incidence of color blindness, namely that it affects 8% of men and only 0.5% of women.

Color blindness probably made life especially difficult for Dalton, as Quakers are traditionally expected to wear simple, plain-colored clothing. There are plenty of wild stories about this, but it is doubtful if they are all true. Dalton is, for example, alleged to have bought his mother bright red stockings for her birthday, thinking that they were a suitable gray color. His mother explained that, as a Quaker, she could never wear such a bright color. Dalton could not believe it and showed the stockings to his brother, who confirmed that they were indeed gray. It was only after several other women supported his mother's side of the story that Dalton realized that he – and his brother – saw colors differently to others. Many years later, during the ceremonial presentation of an honorary doctorate at the University of Edinburgh, he shocked the Anglican bishops present – again unintentionally – with his scarlet clothing. The bishops had expected somewhat less exuberant clothing from a scientist, and certainly from a Quaker.

Strictly speaking, people had known about color blindness long before 1794, when Dalton described how it affected him. Plato spoke of learning problems related to how people see colors, while, in 1686, Richard Waller described the principle of the three primary colors in his small color atlas. An anonymous document on painting miniatures from 1708 describes for the first time how the primary colors – red, yellow, and blue – can be mixed to make all other colors. In 1781, a German journal published an article by an obscure figure by the name of Giros von Gentilly, who described how the retina has three different kinds of molecules or membranes, corresponding to red, green, and blue light. Color blindness would then be caused by one, two, or all three of these molecule types being paralyzed or, conversely, overactive.

Although Dalton was clearly not the first to address the problem, his name is connected with it. His extremely detailed and systematic description of his own color blindness and his later fame ensured that "Daltonism" came to be used as synonym for the defect.

Although atomic theory and color blindness have little in common substantively, there are clear similarities in the way Dalton approached both topics. He was a very independent thinker, writing: "Having been in my progress so often misled by taking for granted the results of others, I have determined to write as little as possible but what I can attest by my own experience."

Although this independent standpoint helped to stimulate Dalton's capacity to think creatively, his unwillingness to embrace others' ideas did hamper his development. He continued, for example, to believe in the validity of his own graphic presentation of the elements, while others were already applying the improved system devised by Jöns Jakob Berzelius.

Dalton's theory of the cause of color blindness also lost credibility during his lifetime, but he remained convinced that his own color blindness was caused by the blue discoloration of his aqueous humor. He was so convinced he was right that he

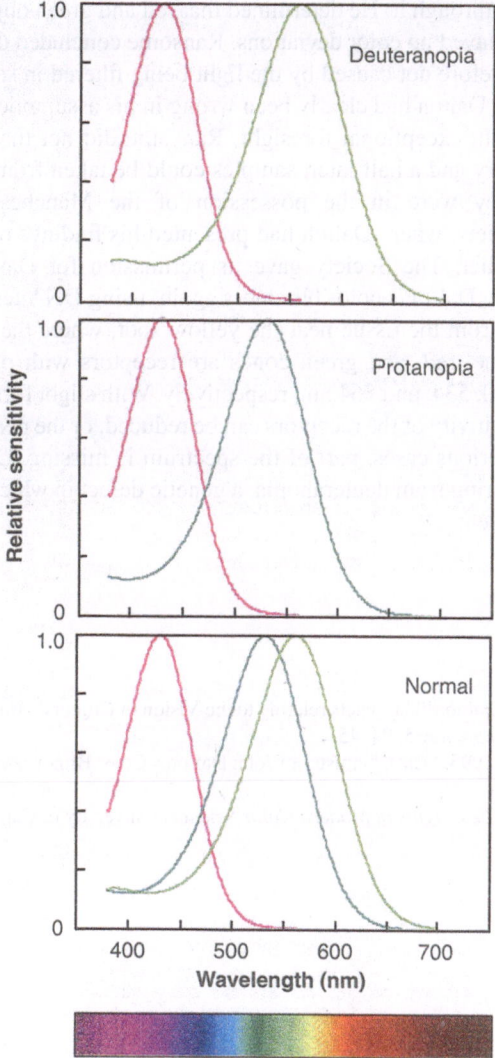

The relative spectral sensitivity of the human eye in normally functioning eyes, protanopia and deuteranopia. With protanopia, there is no sensitivity for the right end of the spectrum, so that red cannot be observed. Deuteranopia sufferers, like Dalton, lack sensitivity to the middle part of the spectrum, so that they are unable to see green.

instructed his physician, Joseph Ransome, to study his eyes immediately after his death to test his hypothesis.

John Dalton died on July 27, 1844. The following day, Ransome collected a little aqueous humor from one of his eyes in a watch glass and observed that it was as clear as normal. Ransome then removed a small layer from the back of the other eye, so that he could look through it. He determined that red and green objects seen through Dalton's eyes displayed no color deviations. Ransome concluded that Dalton's color blindness was therefore not caused by the light being filtered in some way before it reached the retina. Dalton had clearly been wrong in his assumptions.

Fortunately, with exceptional foresight, Ransome did not throw Dalton's eyes away and, a century and a half later, samples could be taken from them for further investigation. They were in the possession of the Manchester Literary and Philosophical Society, where Dalton had presented his findings on color blindness two centuries earlier. The Society gave its permission for David Hunt and his colleagues to study Dalton's color blindness again, using DNA tests. The researchers took samples from the tissue near the yellow spot, where the retinal cones are located. These blue, red, and green cones are receptors with peak sensitivity at wavelengths of 420, 534, and 564 nm, respectively. With slight forms of color blindness, the light sensitivity of the receptors can be reduced, or the peak sensitivity may have shifted. In serious cases, part of the spectrum is missing. Dalton was finally diagnosed as suffering from deuteranopia, a genetic defect in which the reception of green colors is absent.

References

John Dalton, 1798. 'Extraordinary Facts relating to the Vision of Colours'. *Memoirs of the Literary and Philosophical Society* 5, 24–45.

David M. Hunt et al., 1995. 'The Chemistry of John Dalton's Color Blindness'. *Science* 267 (5200) 984–988.

J. D. Mollon, 2003. *The Origins of Modern Color Science*. University of Cambridge, 40 pp.

Thomas Young

Thomas Young was not afraid to call
Newton's particle theory of light into
question. His famous "double-split"
experiment showed unequivocally the
wave nature of light. However, in the
race to decipher the Rosetta Stone,
Young lost out to his rival, the Frenchman
Jean-François Champollion.

R. Schils, *How James Watt Invented the Copier: Forgotten Inventions
of Our Great Scientists*, DOI 10.1007/978-1-4614-0860-4_10,
© Springer Science+Business Media, LLC 2012

Light

Thomas Young was one of the last polymaths. As a teenager, he excelled in classical languages and taught himself mathematics and natural sciences. In 1799, after studying medicine, he opened a medical practice in London, where he specialized in the senses, especially sight. He was one of the first to suspect that the eye has three sensors, each designed to observe a specific color. This theory was confirmed almost half a century later by Hermann von Helmholtz.

Young's interest in the eye soon led him to his first experiments with light. Since the seventeenth century, two theories about the nature of light had been developed: the particle theory of Isaac Newton and the wave theory of Christiaan Huygens. Because of his enormous authority, which continued after this death, Newton's particle theory had few opponents until the early nineteenth century.

Young, however, devised an experiment with which he could demonstrate that light moved in waves. He passed the light from a single source through two small holes, that were close together. He projected the light that radiated from the two holes onto a screen. Where the two beams of light overlapped, he observed alternating bands of light and dark. This led him to conclude that the two beams were interfering with each other. The lighter bands must be caused by peaks or troughs that reinforced each other, while the darker bands occurred where the peak of one beam coincided with the trough of the other. Young was also able to deduce the wavelength of various colors from the interference pattern of the light.

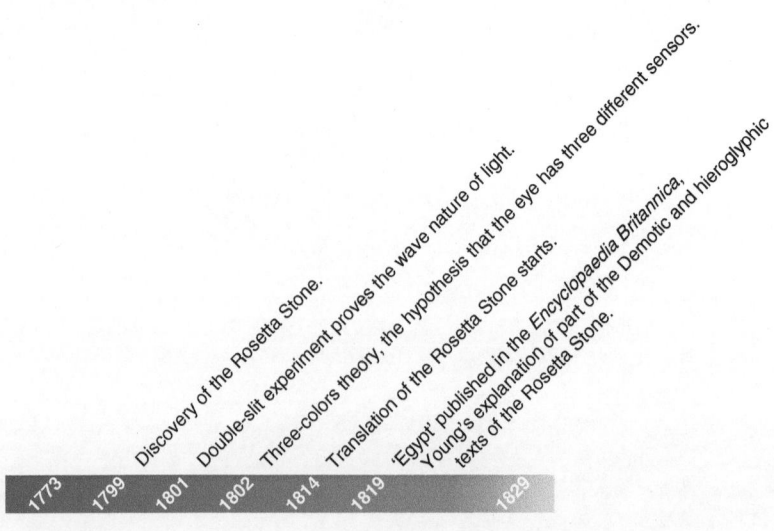

Thomas Young

Other scientists did not easily accept Young's conclusions on the wave nature of light. It was a bold step to refute Newton's claim that light was composed of particles. It was not until Young's contemporary, Augustin-Jean Fresnel, successfully combined the hypotheses of Young and Huygens that the wave nature of light was definitely accepted in Europe.

Rosetta Stone

In May 1798, a French fleet of 400 ships set sail for Egypt. Napoleon believed that, by invading Egypt, he could weaken the links between Great Britain and its colonies in the East. A unique feature of this invasion was that Napeoleon's army was accompanied by a group of 150 scientists, ranging from agricultural experts to mathematicians, and from astronomers to musicologists. The purpose of the scientific mission was to learn more about Egyptian society, culture, and nature. In the years that followed, the scientists' findings were recorded in the comprehensive *Description de l'Egypte*.

Shortly after the French had taken Egypt, the British fleet lay off the coast. To protect the coast from British attack, the French restored Fort St. Julien, near the small town of Rashid (Rosette). During the rebuilding works in the summer of 1799, a French officer stumbled on a piece of granite on which a text was engraved in three languages: Egyptian hieroglyphics, Demotic, and Greek. The stone was about 110 cm high, 80 cm wide, and 30 cm thick. It became clear later that it had originally been considerably taller, and that only a third of the hieroglyphics had been preserved.

It was evident that the Rosetta Stone was a very special find, and it was immediately moved to the headquarters of the French scientists, the *Institut d'Égypte* in Cairo. The stone could be the key to deciphering Egyptian hieroglyphics, knowledge of which had virtually disappeared around the end of the third-century AD.

When the French realized they were losing the battle for Egypt, General Jacques-François Menou had the Stone moved to a safe place. During the capitulation in August 1801, the Rosetta Stone was the subject of a heated discussion between the British General John Hely-Hutchinson and Menou, together with the French scientists. Menou said that the Stone was his own personal property and did not therefore fall under the terms of the surrender. When Hutchinson rejected that argument, the French appealed to the Stone's cultural historical value and accused the British of cultural vandalism. Geoffroy St. Hilaire stated in no uncertain terms that the British would never be able to understand the text on the Stone: "Without us, this material is a dead language, which neither you nor yours can comprehend…" But the French protests were to no avail and, 6 months later, the Stone was in the British Museum in London, where it can still be admired to this day.

In 1814, Thomas Young became acquainted with the Rosetta Stone almost by accident. A friend had brought a Demotic text from the ancient Egyptian city of Thebes, which in the first instance led Young to the Demotic text on the Stone. Young was not a properly trained linguist, but had a talent for languages, especially classical Greek. He also spoke Latin and had some knowledge of Hebrew, Syriac, and Persian.

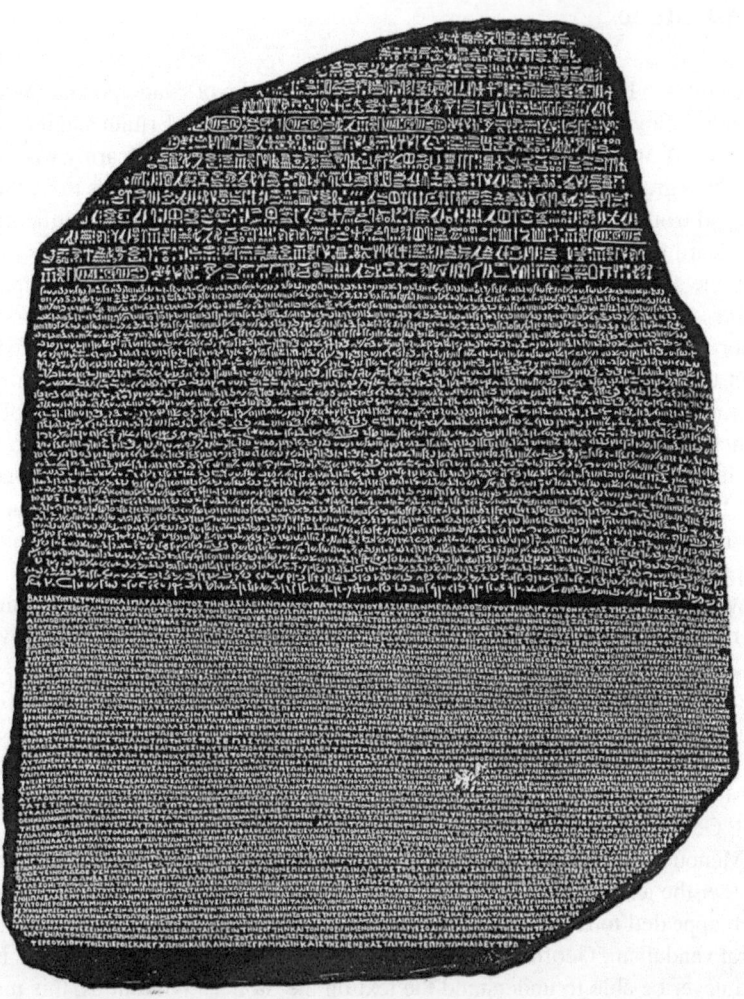

The Rosetta Stone with, from *bottom* to *top*, the Greek, Demotic, and hieroglyphic texts. The text, dated March 27, 196 BC, was a decree to mark the coronation of King Ptolemy V. He is described as a prominent figure, who attached great importance to justice. The decree orders statues of Ptolemy to be erected in the temples and that festivities were to be organized to celebrate the coronation. The author expresses the wish that the text be inscribed on other stones, also in Greek, Demotic, and hieroglyphics

Signes Hiéroglyphiques		Valeur selon Mr Young	Valeur selon mon Alphabet
1		BIR	B
2		E	R
3 ★		I	I . È . AI .
4 ★		N	N
5		inutile	K
6		KE . KEN	S
7		MA	M
8		OLE	L
9 ★		P	P
10		inutile	Ô . OU
11		OS . OSCH .	S
12 ★		'T	T
13		OU	KH . SCH
14 ★		F	F . V
15		ENE	'T

Table from *Precis du Système Hieroglyphiques* by Jean-François Champollion, showing the translation of a number of symbols on the basis of the cartouches of Ptolemy and Berenice. From *left* to *right*, the hieroglyphic text, Young's translation, and Champollion's translation

The Greek part, which was of course understood quite quickly after the Stone was discovered, explained that the text of all three parts was more or less the same. It was therefore logical to use the Greek to translate the two unknown texts.

It was widely believed at the time that hieroglyphics were symbols for words and ideas, while Demotic script was mainly phonetic, with the symbols representing a certain sound. The first thing that Young noticed, however, was the striking resemblance between some of the Demotic symbols and the corresponding hieroglyphics. It looked as though the Demotic text was not a completely different script but was directly related to the hieroglyphics, in the way that a printed letter resembles its handwritten equivalent. Young believed he had found evidence of how the hieroglyphic symbols of human forms, animals, plants, and other objects had developed into handwritten Demotic signs. He came to the conclusion that Demotic signs did not constitute an alphabet but were imitations of hieroglyphics, mixed with letters of the alphabet.

This was a very significant discovery, but Young did not dare to take the next step: to cast doubt on the assumption that the hieroglyphics formed a purely symbolic script. It was the Frenchman Jean-François Champollion who was later to take this groundbreaking step forward.

Earlier researchers thought that the only phonetic elements in hieroglyphics were the names of kings and queens, which are surrounded by oval rings known as cartouches. Young, too, tried to analyze the cartouches. The Greek translation showed that the text should contain cartouches of King Ptolemy and Queen Berenice. Using these two names, Young was able to determine the phonetic meaning of a small number of hieroglyphics. Of the symbols in the two cartouches, he interpreted six correctly and three partly correctly, but four were definitely wrong.

Young published his findings in 1819 in the article on Egypt in the fourth edition of the *Encyclopaedia Britannica*. He presented the meanings of 218 Demotic and 200 hieroglyphic words, 80 of which later proved correct.

Strangely enough, he then gave up working on the texts. It never became clear why, as he never revealed his reasons for doing so. Perhaps he simply lost interest. After all it was not the first time Young's groundbreaking discoveries had been developed further by others. It was Fresnel who finally provided confirmation of the wave nature of light, while Young's three color theory was later worked out in detail by Helmholtz. In this case, it was Champollion who took over the baton.

Champollion had been fascinated by Egyptian culture since his youth. From 1814, armed with a thorough background in languages, he set about deciphering the hieroglyphic script. However, he had to use copies, which were not always as clear as the original and, at first, he made little progress. The turning point, however, came in 1821 when he was studying a bilingual text on an obelisk, which included the name of Cleopatra. While analyzing the text, Champollion had the revolutionary idea that hieroglyphics were a combined script. In 1824, he published *Précis du Système Hieroglyphiques* (Concise Overview of the Hieroglyphic System), a complete description of hieroglyphic script, the sacred language used primarily in religious and aristocratic circles. Young's suspicion that Demotic script, the language of the common man, was derived from it, proved to be correct.

There is no evidence to suggest that, at first, Young and Champollion's interest in the Stone was anything other than scientific curiosity. It may be coincidence that they both decided to try and decipher it in the same year. But once the work started in earnest, a personal rivalry developed in which national honor was at stake. Champollion wrote to his brother: "The Brit can do whatever he wants – it will remain ours: and all of old England will learn from young France how to spell hieroglyphs using an entirely different method ..."

The Rosetta Stone is still a source of deep umbrage between the two nations, as the following incident shows. To mark the 150th anniversary of the decipherment of the Stone, it was sent to the Louvre in Paris on temporary loan. Portraits of Champollion and Young were carefully selected and were of identical size. Nevertheless, French visitors complained that the picture of Young was larger, while their British counterparts were convinced that the reverse was true.

References

Thomas Young, 1823. *An Account of Some Recent Discoveries in Hieroglyphical Literature and Egyptian Antiquities*. John Murray, 194 pp.

Muriel Mirak Weissbach, 2000. 'Jean François Champollion and the True Story of Egypt'. *21st Century Science & Technology magazine* 12 (4), 26–39.

Andrew Robinson, 2007. 'Thomas Young and the Rosetta Stone'. *Endeavour* 31 (2), 59–64.

John Ray, 2007. *The Rosetta Stone and the Rebirth of Ancient Egypt*. Harvard University Press, 208 pp.

Justus von Liebig

Justus von Liebig is best known for inventing fertilizer. His strong desire to feed the world also led him to develop the "poor man's meat," ready-made bouillon.

R. Schils, *How James Watt Invented the Copier: Forgotten Inventions of Our Great Scientists*, DOI 10.1007/978-1-4614-0860-4_11,
© Springer Science+Business Media, LLC 2012

Fertilizer

Justus von Liebig was familiar with experiments from an early age. His father traded in chemicals and made all kinds of paint and varnish himself. In his father's workshop, Justus developed his own chemical knowledge largely by conducting his own experiments and filling in the gaps by reading as much as possible on chemistry in his local library. Although he did not complete secondary school, by the age of 21, he was a professor in chemistry at the University of Giessen.

At the university, Liebig put the chemical sciences firmly on the map, first inorganic and later organic chemistry. The organization and equipment of his laboratory in particular became an example for others throughout Europe. By the end of the 1830s, Liebig had become a highly valued and renowned chemist with more than 300 scientific publications to his name.

However, Liebig was more than just a fundamental scientist; he considered it of great importance that his newly acquired knowledge was applied in practice. In 1840, he published the agricultural classic *Die organische Chemie in ihrer Anwendung auf Agricultur und Physiologie* (*Organic Chemistry in its Application to Agriculture and Physiology*), in which he shows that plants feed on simple mineral elements and salts

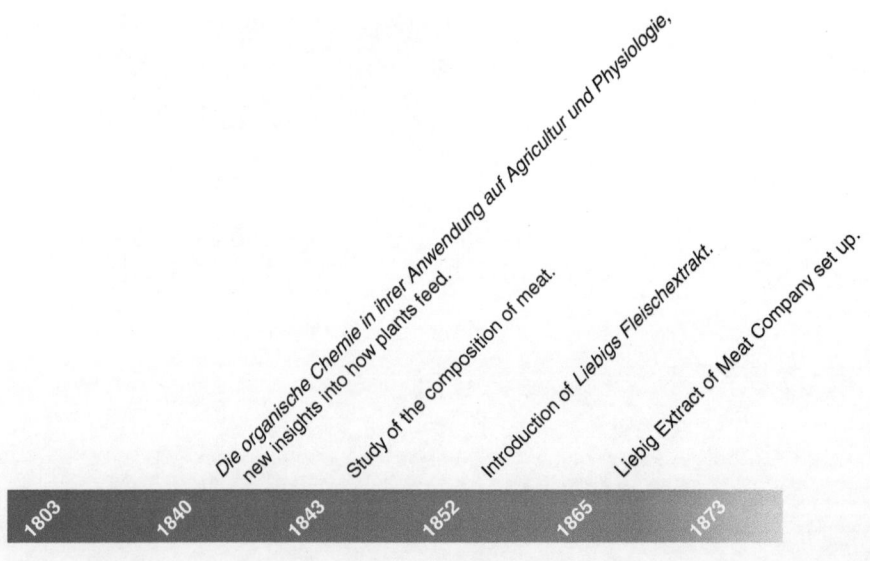

Justus von Liebig

dissolved in water. This directly conflicts with the "humus theory," which states that plant growth depends exclusively on the organic content of the soil. He also shows that the growth of a plant is determined by the nutrient that is in the shortest supply.

Liebig's Law of the Minimum implies that, if the supply of the nutrient that is available in insufficient quantities is not increased, adding other nutrients is completely futile. Liebig subsequently developed the first mixtures of mineral salts to compensate for shortages of essential nutrients. With these first fertilizers, Liebig helped initiate the green revolution in European agriculture, which led to an explosive increase in the production of food crops.

Liebig did not, however, limit himself to plant nutrition, but looked further, investigating the underlying principles of animal nutrition, which he described in *Die Tierchemie oder die organische Chemie in ihrer Anwendung auf Physiologie und Pathologie* (*Organic Chemistry in its Application to Physiology and Pathology*), published in 1842.

More than a 100 years later, it emerged that the honor of initiating the agricultural revolution could not be attributed solely to Liebig. His work proved not to be entirely original. His compatriot Carl Sprengel had apparently established, some years before Liebig, that plants feed on mineral salts. Sprengel had also been the first to formulate the Law of the Minimum. However, Sprengel's publications had never attracted much attention, while Liebig, who enjoyed greater scientific renown, was in a better position to act as advocate for the new science of plant nutrition. In 1995, to compensate somewhat for this oversight, the Sprengel–Liebig medal was introduced in Germany for individuals who have made an exceptional contribution to agriculture.

Bouillon

In the mid-nineteenth century, Europe was completely in the grip of the Industrial Revolution. With the growing demand for labor, the cities were inundated with newcomers, increasing demand for food to feed the growing masses. At that time, meat was the main source of protein and a staple component of the daily menu, yet large parts of the population could hardly afford it.

Justus von Liebig sought to solve this problem by producing nutritious meat extracts. From 1843, together with this students, he studied the composition of the meat from different animals. He developed a method of extracting the proteins and other nutrients from meat. It entailed simmering finely minced beef in hot – but not boiling – water. The water slowly evaporated, leaving a soft, brown extract, rich in protein and full of flavor, that made a delicious and nutritious bouillon when boiling water was added.

In the years that followed, the meat extract was produced on a small scale by the Royal pharmacist Max von Pettenkofer, at first under the name *Extractum Carnis* and, from 1852, as *Liebigs Fleischextrakt* (Liebig's Meat Extract). It was largely sold as a food supplement and medicinal remedy.

No one less than Florence Nightingale praised *Liebigs Fleischextract* for its heal-ing powers. She used the extract during the Crimean War to help sick and wounded soldiers to recover. The French troops also used the miracle cure, but with a

Sales of *Liebigs Fleischextrakt* were promoted with a clever marketing campaign. With every purchase, the buyer was given a beautifully illustrated card. Between 1870 and 1975, around 11,000 different cards were distributed

customary culinary modification: badly wounded soldiers were administered with a solution containing one part extract and eight parts wine. This was considered sufficient for them to survive being moved to the nearest field hospital.

Despite all the extract's favorable properties, Liebig was unable to get it produced commercially on a large scale. He needed 30 kg of meat for 1 kg of extract, making it too expensive as food for the common man. But he did not give up, and published his findings in the 32nd edition of his *Chemische Briefe* (Chemical Letters). He himself found the taste of the extract unsurpassed: "The taste of the dried meat extract is of such intensity, no other kitchen aid has a comparable herbal strength." In the letters, he also described the possibilities he saw for countries like Argentina and Australia, where he has heard they had meat in abundance. He promised to share his idea with anyone who was in a position to produce the extract on a large scale. For the time being, however, there were no takers and the plans gathered dust on the shelf.

Some years after Liebig's publication, engineer Georg Christian Giebert was traveling through South America. In many places in Uruguay, he saw unused cattle carcasses lying around. He discovered that the animals had been bred and slaughtered solely for their hides, horns, and fat. He realized that he had found a solution to Liebig's problem – affordable meat – and, in 1861, traveled to Germany. He met

Part of the production plant in Uruguay, some years before the Liebig Extract of Meat Company moved production to other locations

Liebig and proposed producing meat extract on a large scale in Uruguay for export to Europe. Liebig was immediately enthusiastic, but had doubts about the quality of the extract. He was only prepared to lend his name to the product if Pettenkofer's pharmacy could continually monitor the quality. Giebert agreed to the conditions and, after being taught how to make the extract, returned to Uruguay to start production.

Giebert succeeded in producing the extract for a third of the costs in Europe. In November 1862, he sent the first samples to Liebig, who responded very enthusiastically: "I can be satisfied that the quality of the samples is much better than I expected, because the meat comes from practically wild animals." Liebig and Giebert were both now so enthusiastic that they decided to produce the meat extract on a large scale.

In 1865, the Liebig Extract of Meat Company was formally founded, producing beef extract as a cheap and nutritious alternative to "real" meat. Liebig became a director and was in charge of the scientific department in Munich, where the end product was checked and analyzed.

Production increased spectacularly, from 28 tons in 1865 to 421 tons in 1871. The extract was packaged for the shops in glass jars. In Europe, it was no longer only recommended as a food supplement and remedy, but was also increasingly used as a "normal" food.

Liebigs Fleischextrakt cannot be compared to the ready-to-use bouillon cubes we use nowadays. The step from the liquid meat extract to the bouillon cube was not made until the end of the nineteenth century, by Swiss entrepreneur Julius Maggi. Maggi first experimented with a method of making a basic soup by grinding cheap and nutritious pulses. Later, he also used meat extract, drying it, mixing it with herbs, shredded vegetables and salt, and pressing it into cubes. Today's bouillon cubes are no longer considered a source of protein, as we now obtain our protein from a wide variety of other foods.

By 1875, Giebert and Liebig were both dead, but the company was still in full production. Every year, the plant processed some 150,000 beef cattle and exported 500 tons of extract to Europe. Besides the extract, the company produced other meat products, including corned beef, tongue, tallow, and fertilizer. The company became a driver of the Uruguayan cattle farming sector and was seen as a textbook example of an efficient meat processing plant. Nothing of the animal was wasted, every part being used one way or another.

In 1924, the Liebig Extract of Meat Company in Uruguay ceased operations. Production, now past its peak, was moved to other parts of the world. In the second half of the twentieth century, the company was involved in a series of takeovers and mergers, and became part of other meat-processing concerns. The sales of *Liebigs Fleischextract* fell in the face of competition from cheaper alternatives from Maggi and Oxo. Yet it can still be bought today in delicatessen shops.

The plant in Uruguay came to a less fortunate end. After the departure of the Liebig Extract of Meat Company, it continued to produce corned beef and other meat products, as part of a different company. At its peak, in 1964, the plant employed 64,000 people. In that year, an outbreak of typhoid in Aberdeen was traced back to the plant, heralding the beginning of the end. In 1971, it was donated to the Uruguayan government and, 8 years later, it closed its gates for good. All that remains now is an industrial monument.

References

Justus von Liebig, 1878. *Chemische Briefe*, sixth edition. Leipzig und Heidelberg. C.F. Winter'sche Verlagshandlung.

William Brock, 2002. *Justus Von Liebig: The Chemical Gatekeeper*. Cambridge University Press, 392 pp.

Günther Klaus Judel, 2003. 'Die Geschichte von Liebigs Fleischextrakt'. *Spiegel der Forschung* 20 (1), 6–17.

Charles Darwin

Charles Darwin is the founder of the theory of evolution, but he also won his spurs as a pedologist. He started and ended his career with publications on earthworms.

R. Schils, *How James Watt Invented the Copier: Forgotten Inventions of Our Great Scientists*, DOI 10.1007/978-1-4614-0860-4_12,
© Springer Science+Business Media, LLC 2012

Theory of Evolution

Charles Darwin went to the University of Edinburgh where, at a young age, he came into contact with students and teachers who were not afraid to hold nonconformist views. The university was a popular alternative for dissident students who had been refused admittance to Oxford or Cambridge. Robert Edmond Grant, a radical supporter of French biologist Jean-Baptiste Lamarck, became Darwin's mentor. At the beginning of the nineteenth century, Grant had already proposed that marine invertebrates may have played an important role in the development of more complex life forms, an assumption that could not count on much support at that time. After Darwin had been at Edinburgh for 3 years, his father decided that enough was enough and that Cambridge was after all a better environment for his son. In Cambridge, Darwin became acquainted with the more classical approach to animal and plant sciences.

In 1831, he was given the opportunity to travel to South America on board the HMS Beagle, in the first instance not as a scientist, but as a table companion for the captain. The voyage largely followed the coastline of South America, but also included countries like New Zealand and Australia. Darwin observed and collected countless species of plants and animals. Fascinated by the geographical distribution of living plants and animals, but also by the many fossils he discovered, he began to study the way species gradually changed.

Back in England, Darwin developed a theory that describes the origin of species by a form of natural selection. He found an important piece of the puzzle in the *Principle of Population* by economist Thomas Malthus, which describes the link

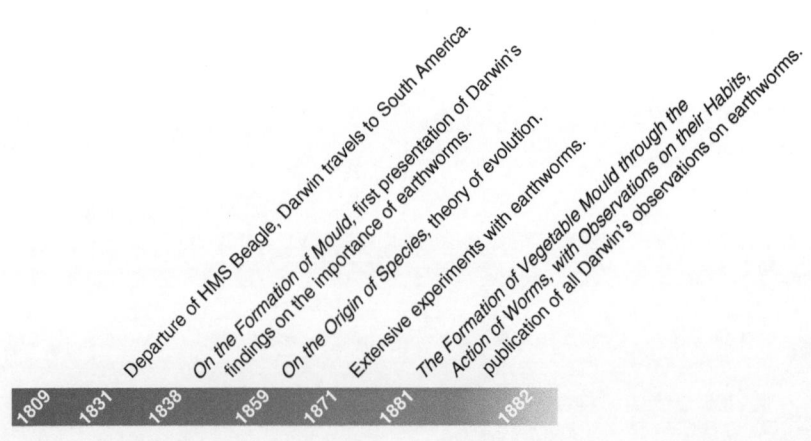

Charles Darwin

between population growth and food scarcity. Darwin realized that an explosive increase in animal population also leads to a shortage of food and that, in the ensuing competition, the weaker individuals are eradicated. Although his theory of natural selection was already complete in 1840, it was a long time before he had sufficient evidence to convince his critical and suspicious contemporaries. After all, the suggestion that animals and people had common predecessors was a shocking claim in Victorian England. In 1859, however, Darwin finally published his masterpiece, *On the Origin of Species by Means of Natural Selection, or the Preservation of Favoured Races in the Struggle for Life.*

Illustration on the cover of the satirical magazine *Punch* on December 6, 1881, around 2 months after publication of Darwin's book on worms

Earthworms

Darwin's first serious encounter with worms came in 1837, when he had just returned from South America. Although he was satisfied with the impressive collection of plants and animals he had brought back, the long voyage had exhausted him. At first, he made good progress with writing up his journal, which extended to thousands of pages, but the exhaustion soon took its toll. On the advice of his friends, Darwin decided to take a rest in the country for several weeks and went to visit his uncle, Josiah Wedgwood.

He did not, however, have much opportunity to rest. His uncle took him to three meadows where, many years previously, farmers had spread lime, marl, and cinders. These fragments were now covered by a layer of soil several centimeters thick. Wedgwood was convinced that this had been caused by the activity of earthworms and not, as the farmers thought, by downward movement of the fragments themselves.

Darwin was immediately fascinated by his uncle's observation. He knew better than anyone that the enormous volume of soil that earthworms could move played a significant role in soil formation. Shortly afterwards, he presented his findings on worms to the Geological Society of London. But Darwin's argument did not convince his audience. How could a serious scientist, at a time when the most exotic animal species were being discovered all around the world, spend his valuable time on such a small and insignificant creature as the earthworm? One critic commented:

Drawing of cross-section of soil to a depth of around 13 cm, from October 1837. This plot of meadow had been plowed, harrowed, and covered with a layer of burned marl and cinders 15 years earlier

"In the eyes of most men ... the earthworm is a mere blind, dumb, senseless, and unpleasantly slimy annelid. Mr. Darwin undertakes to rehabilitate his character, and the earthworm steps forth at once as an intelligent and beneficent personage, a worker of vast geological changes, a planer down of mountainsides ... a friend of man."

But Darwin refused to be deterred by the criticism. "The subject may appear an insignificant one," he said, "but we shall see that it possesses some interest." Darwin found that his critics lacked the capacity to visualize the long-term effects of small, recurring processes, which he felt hampered the advance of science.

Darwin shifted his attention to other topics, but certainly retained his interest in earthworms, as is shown by short articles he published in the *Gardener's Chronicle and Agricultural Gazette* in 1844 and 1869.

In 1871, he returned to the subject in earnest, corresponding extensively with colleagues. More importantly, he also conducted a series of detailed experiments. He set to work with the help of his family, and especially his three sons. For a decade, he once again devoted his full attention to worms, observing their behavior closely. He exposed them to a range of stimuli, including touch, voices, and piano music, and odors of varying intensity, including tobacco smoke. He studied how worms feed, offering them fat, raw meat, lettuce, onions, and starch. He observed that they tugged leaves into their burrow systems, for food and as insulation against excessive fluctuations in temperature and humidity. He was particularly fascinated by the way in which the worms dragged the leaves into their burrows, concluding that they did not do it randomly, but learned from experience what worked best.

Darwin devised a number of experiments to test his hypotheses on the intelligence of worms. He offered them leaves, stalks, twigs, and pine needles in a wide variety of shapes and sizes. To exclude the possibility that they were calling on previous experience, he tried offering leaves from exotic plants. He even went as far as to cut small pieces of paper into different triangular shapes. He then observed how the worms dragged the different materials into the ground. They proved indeed to work according to a fixed pattern. This led Darwin to conclude that worms must have at least some form of intelligence, as they acted in a similar way to how people would respond in comparable circumstances. Darwin's conclusion remains controversial to the present day.

This was, however, not Darwin's only achievement in his study of worms. He proposed an initial estimate of the number of worms in the ground, at around 130,000 per hectare. Today, we know that the number of worms per hectare can vary from a few tens of thousands to as many as ten million, depending on the sort of soil and vegetation, temperature, humidity, fertilizer use, and acidity. Darwin also studied the part played by worms in stone and soil erosion. He was one of the few scientists to understand that the accumulation of a large number of small, apparently insignificant, events can lead to major changes in the long term. And there lies a striking similarity between evolution and the work of earthworms.

Despite all the energy that Darwin invested in his book on worms, his expectations were modest. In September 1880, he wrote to Victor Carus: "I am writing a very little book, ... As far as I can judge it will be a curious little book." In 1881, a year before his death, Darwin published his findings in *The Formation of Vegetable Mould Through the Action of Worms, with Observations on Their Habits.*

As had happened 40 years earlier with his first presentation on worms, the book was again unfavorably received in scientific circles. At that time, worms were primarily seen as a pest, devouring the roots of plants and disfiguring perfect lawns with small piles of excrement, and the *Complete Course of Agriculture* presented various methods of combating earthworms. Darwin's book appeared at a time when agricultural research was focusing on the chemical properties of the soil, as demonstrated by the work by Justus von Liebig published 40 years earlier. At that time, no one was prepared to accept that worms were important for crop production. With the advent of biological farming methods in the twentieth century, there was renewed interest in Darwin's book on worms. And today, soil life is experiencing an impressive revival of interest in conventional agriculture.

Darwin could, however, find solace in one thing: the general public was very enthusiastic about his book, probably because of his interesting writing style and his appealing conclusions about the intelligence of worms. Darwin himself was very surprised by the great public interest in his book. On November 5, 1881, publisher John Murray wrote to him: "We have now sold 3,500 worms!!!" Three years later, that had risen to 8,500, a success comparable to his other bestseller, *On the Origin of Species.*

References

Charles Darwin, 1838. On the Formation of Mould. *Proceedings of the Geological Society of London* 2, 574–576.

Charles Darwin, 1881. *The Formation of Vegetable Mould Through the Action of Worms, with Observations on their Habits.* John Murray, London.

George G. Brown et al., 2002. 'With Darwin, earthworms turn intelligent and become human friends'. *Pedo biologica* 47, 924–933.

Christian Feller et al., 2003. 'Charles Darwin, earthworms and the natural sciences: various lessons from past to future'. *Agriculture Ecosystems & Environment* 99, 29–49.

William Thomson (Lord Kelvin)

The absolute temperature scale was posthumously named after Lord Kelvin for his contribution to thermo-dynamics. Lord Kelvin, born as William Thomson, also played a crucial role in the laying of the first transatlantic telegraph cables.

R. Schils, *How James Watt Invented the Copier: Forgotten Inventions of Our Great Scientists*, DOI 10.1007/978-1-4614-0860-4_13,
© Springer Science+Business Media, LLC 2012

Absolute Temperature

In 1840, William Thomson read Joseph Fourier's *Théorie analytique de la chaleur* (Analytical theory of heat), a groundbreaking work on heat, and he remained fascinated by the subject for the rest of his life. In one of his first papers, Thomson elaborated on Fourier's conclusion that, in the course of time, heat in a body will always be distributed evenly. Thomson reversed the argument, stating that a body with a uniform temperature cannot be calculated endlessly back in time as that would create mathematically impossible temperature distributions. He applied this argument to the Earth, concluding that it cannot be infinitely old. His estimates of the age of the planet ranged from 20 to 400 million years, which evolutionists and geologists considered far too low. As the evidence that the Earth was much older gradually accumulated, Thomson stubbornly stuck to his own views. He would have found today's estimate of 4.5 billion years absurdly high.

Scientific interest in heat partly had its origins in the advance of the steam engine in the nineteenth century. It was remarkable that, while one practical improvement followed the other, fundamental understanding of the relation between heat and work lagged far behind. A first requirement for studying heat was a reliable and uniform method for measuring temperature. Although good thermometers were available, what Thomson felt was lacking was a theoretical principle for an absolute temperature scale.

He sought a solution in the pioneering work of Nicolas Léonard Sadi Carnot on heat and work. Carnot had devised a hypothetical model to determine the efficiency of steam engines. The "Carnot cycle" occurred in an idealized engine, in which a gas was alternately heated and cooled in a cylinder with a movable piston. According to Carnot, the work performed by the heat was directly related to the difference

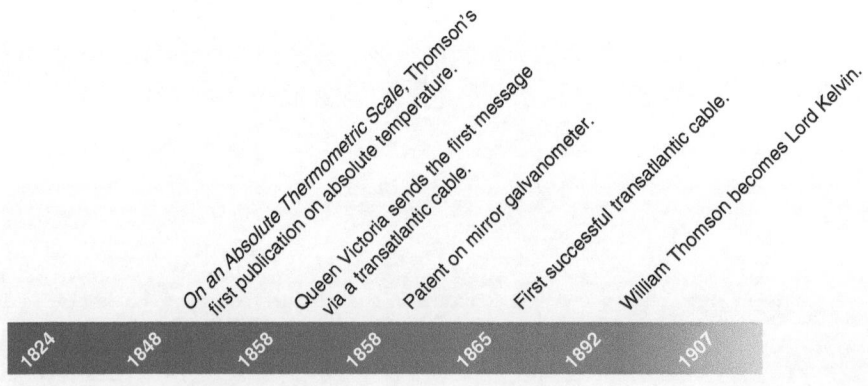

William Thomson (Lord Kelvin)

between the lowest and highest temperatures in the cycle. That was exactly what Thomson was looking for, an independent measure of absolute temperature differences.

In a series of papers published from 1848 onward, Thomson linked his definition of absolute temperature to the Carnot cycle, in the first instance to the work produced, and secondly to the ratio between the heat absorbed and the heat emitted. Thomson now had his theoretical definition, but because the Carnot cycle could not be reproduced in reality, he still had to use existing thermometers. As with the Celsius scale, Thomson divided the difference between the freezing point and the boiling point of water into a scale of a hundred degrees. From the work of James Joule, he used 273.7 as freezing point and 373.7 as boiling point, very close to the current values of 273.15 and 373.15.

Today, we know that the absolute minimum temperature is −273.16°C, which Thomson and others had deduced from the ideal gas law, to within a few decimal points. But, at that point, they had no idea of the physical meaning of the minimum temperature and saw it more as a kind of calibration point for thermometers. At a later stage of his quest for the absolute temperature scale, Thomson proposed that the efficiency of a Carnot engine would be practically 100% if the absolute temperature were zero. Lower temperatures would therefore have no significance.

In 1892, William Thomson was raised to the peerage by Queen Victoria and chose the name Kelvin, after a small River in Glasgow. Since 1954, the kelvin has been the official unit for the absolute temperature.

Transatlantic Cable

In the early nineteenth century, electrical telegraphy unleashed a veritable revolution in communication. An extensive network of telegraph cables soon emerged throughout Europe and the USA. However, there was still no connection between the two continents.

From 1838, the first cautious attempts were made to lay a cable under water, but insulating the copper wiring was a serious problem. The first cables were wrapped in cloth, soaked in tar. This proved effective over short distances, through rivers and ports, but was unfeasible for longer distances. Crossing the Atlantic Ocean was still completely out of the question. That changed in 1848, when gutta-percha became available. Gutta-percha is a rubber-like substance extracted from the sap of the gutta-percha tree (*Palaquium gutta*). It proved an excellent insulator that was easy to work with when heated. At low temperatures it is solid, yet flexible. Gutta-percha was used to insulate telegraph cables across the Irish Sea and across the English Channel between Dover and Calais.

It was now only a small step to laying a successful transatlantic cable, but there was another unexpected problem. Underwater cables proved to conduct electrical signals less efficiently than those above ground. Instead of being clear and short, the signal was vague and drawn out. The problem was presented to Michael Faraday, who sought the cause of the interference in the high conductivity of the surrounding water. He concluded that, for such great distances, the cable would have to be, as it were, charged, like a kind of capacitor. But, as was often the case with Faraday, he left it there, and went no further than a qualitative analysis.

William Thomson heard about the problem with the signal indirectly, and he quickly found a solution where Faraday had failed. He wrote a quantitative theoretical analysis on the transfer of an electrical signal through an insulated submarine cable. He calculated that the front of the signal had a constant speed, but that the arrival time of the peak of the wave increased by the square of the distance covered. He concluded that, to maintain the quality of the signal, it was necessary to use a cable with a larger diameter. Technically and economically, of course, that was not such good news, but Thomson responded coolly, saying that it was a simple calculation: the optimal thickness of the core and the insulation could be easily calculated for each desired transfer speed on the basis of the price of copper and gutta-percha.

His earlier work on heat had taught Thomson that scientific knowledge was rarely applied in practice. What applied to steam engines, was not much different in the world of telegraphy. The great pioneers clearly found Thomson's analysis a step too far.

One of them was Edward Orange Wildman Whitehouse, who was responsible for technical matters at the Atlantic Telegraph Company. He did not agree with Thomson, as his own observations suggested that transfer time was linearly related to the length of the cable and not to the square of the length. This was an important difference as, in his view, it made the Atlantic crossing less problematic than Thomson thought.

The owner of the Atlantic Telegraph Company, Cyrus Field, was in a hurry to lay a transatlantic cable. Despite the conflicting advice of Whitehouse and Thomson, he ordered 4,000 km of cable for a link between Newfoundland and Ireland. He clearly trusted Whitehouse's practical approach more than Thomson's scientific perspective.

In August 1857, two ships left Ireland with the first transatlantic cable on board. It was a festive departure, with Field reading a message from American President

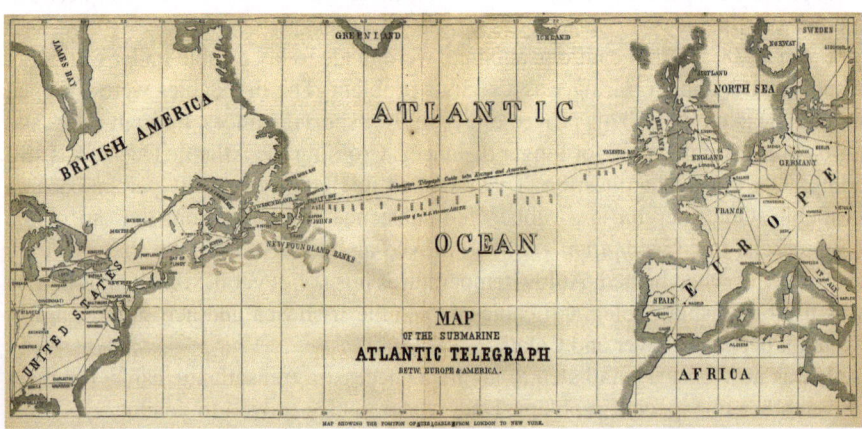

Map showing the position of the first transatlantic cable

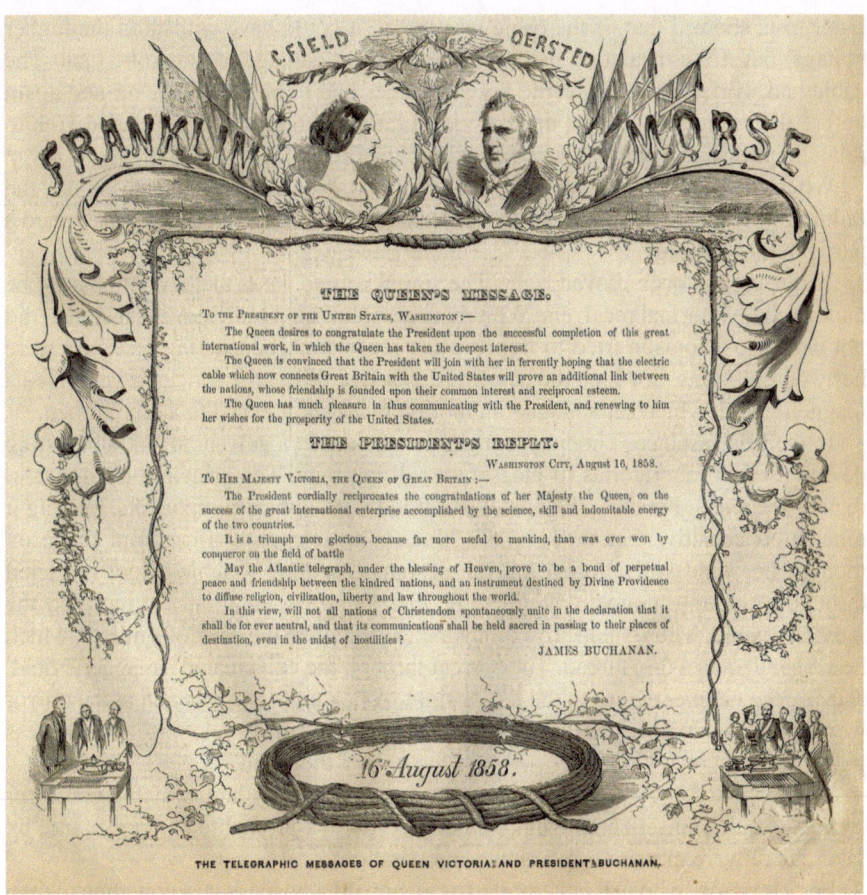

The first transatlantic messages between Queen Victoria and President Buchanan

James Buchanan inviting Queen Victoria to send the first message through the cable. But the euphoria did not last long. Within a few days, the fairytale came to an end. After an unfortunate break, 500 km of cable lay useless on the seabed.

A second attempt, a year later, was successful. On August 5, 1858, the first transatlantic cable was completed, to an uproarious reception from the press. But after a few days, the enthusiasm died down, as the message from the Queen did not materialize. Eleven days later, to widespread relief, it was announced that the Queen's message had finally been received. In the weeks that followed, hundreds of messages were sent through the cable, but shortly after, it fell silent for good.

Although the exact cause of the malfunction has never been discovered, it emerged that Whitehouse had used brute force to try and "pump" the signal through the cable. As normal telegraph receivers could not pick up the weak signal from the transatlantic cable, Whitehouse tried to strengthen it by increasing the electrical potential. Using a transformer, he cranked the voltage up to around 2,000 volts.

Later tests showed that, if the cable was intact, it could have withstood the higher voltage, but if the insulation had even the slightest damage, it would be fatal. The cable had worked, but with difficulty. Messages had to be repeated over and again before their content could be understood. It later emerged that it had taken 16 h to relay the message from the Queen clearly to the other side.

When the truth became known, Whitehouse was dismissed. For the British, the failure of the prestigious project was so serious that, in 1859, they even instigated a parliamentary inquiry.

Thomson had been proven right. The signal was so weak that it could hardly be understood by normal receivers. Whitehouse's solution, a more powerful signal, did not work, so Thomson tried to improve the receivers. The existing receivers were hefty contraptions, in which the current was passed through a coil. This generated a magnetic field, which set a magnet in motion that was attached to a kind of pen.

In the first instance, Thomson sought the answer in a galvanometer, an existing device used in laboratories to measure small currents. The principle is similar to regular receivers, but the galvanometer is fitted with lighter components, making it much more sensitive than the hefty machines in the telegraph offices. But Thomson thought he could still improve on the design. One day, watching light refracted through a rotating monocle, he knew he had found the solution. He replaced the moving magnet with a small strip of magnetized steel on the back of a mirror, which he suspended on a thin thread. The current through the coil created a magnetic field, causing the magnet to turn to the left or right. A beam of light directed at the mirror would then move along a scale. Thomson succeeded in developing a weightless pointer, so that reception increased by a factor of a thousand.

In 1865 and 1866, two transatlantic cables were finally laid, which worked for some 5 years. Transatlantic telegraphy proved to be a profitable enterprise and, by 1865, there were more than ten cables in operation.

For Thomson, who had once started as an unpaid advisor, it was now time to reap the fruits of his ideas. His primary source of income was the patent rights on the mirror galvanometer. He also advised on many other cable projects, but now for a fee. In 1892, William Thomson was made a peer, not as much for his scientific achievements as for his commercial success and personal wealth. He was the text-book example of a successful Victorian entrepreneur, whose products contributed to the imperial ambitions of the UK.

References

William Thomson, 1855. 'On the Theory of the Electric Telegraph'. *Proceedings of the Royal Society of London* 7, 382–399.
William Thomson, 1856. 'On Practical Methods for Rapid Signalling by the Electric Telegraph'. *Proceedings of the Royal Society of London* 8, 299–303.
David Lindley, 2004. *Degrees Kelvin: a Tale of Genius, Invention and Tragedy*. Joseph Henry Press, Washington, 366 pp.

James Clerk Maxwell

James Clerk Maxwell was one of the
greatest physicists of the nineteenth
century. He showed that electricity
and magnetism are not different, but
are part of the same system. Maxwell
also experimented with colors and
produced the first color photograph.

R. Schils, *How James Watt Invented the Copier: Forgotten Inventions
of Our Great Scientists*, DOI 10.1007/978-1-4614-0860-4_14,
© Springer Science+Business Media, LLC 2012

Maxwell's Equations

James Clerk Maxwell was an early developer. He published his first scientific paper at the age of 14, describing a method of drawing Cartesian ovals. He studied natural sciences at Edinburgh, followed by mathematics at Cambridge. After completing his studies, he continued to move regularly between Scotland and England, spending some time as a professor of natural sciences at the Universities of Aberdeen and London. He was awarded his first chair, in experimental physics at Cambridge, in 1871, after spending 4 years at Glenair, his Scottish estate. While he was in Cambridge, he set up the Cavendish Laboratory, where many Nobel Prizewinners, including Ernest Rutherford and James Watson, would later work. Maxwell's most renowned period was undoubtedly his time at King's College, London, in the early 1860s when he brought together existing knowledge on electrical and magnetic phenomena in a single electromagnetic theory.

Magnetic and electrical phenomena had already been described in relative detail in the eighteenth century. It was not, however, until the first half of the nineteenth century that it became clear that the two were linked. Danish physicist Hans Christian Ørsted showed that electrical currents generate magnetic fields while, conversely, Michael Faraday discovered that moving magnetic fields generate electrical currents. Faraday then took a major step forward with his theory on electrical and magnetic force fields. The new theory initially received little support, but Maxwell realized its importance and elaborated on Faraday's ideas.

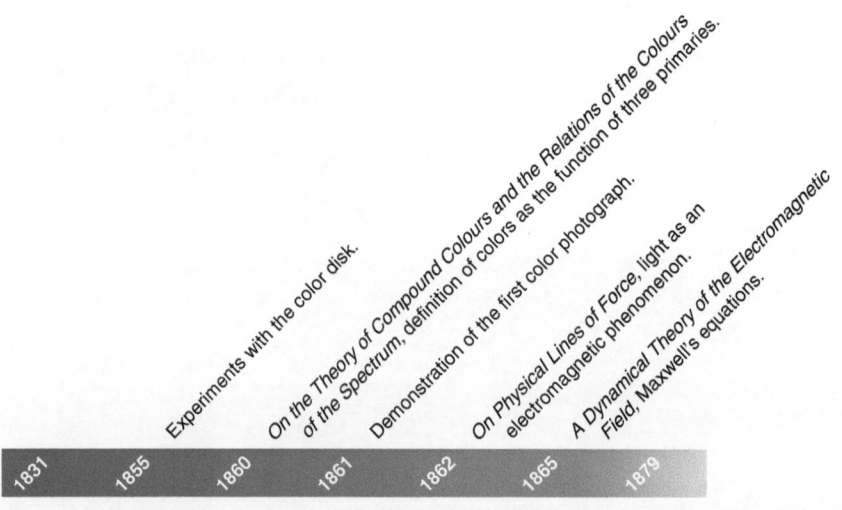

James Clerk Maxwell

Maxwell succeeded in describing Faraday's theory in mathematical terms, which Faraday himself had great difficulty doing. He produced his Maxwell's equations in 1865, beginning with 20 equations with as many variables. Using vector notation, these were later rewritten in a set of four equations. The equations describe how electrical fields are generated by an electrical charge, how electrical currents create magnetic fields, and how electrical fields are generated by changing magnetic fields.

Maxwell suggested that electromagnetic waves can move through space at a speed of 3.1×10^8 m/s. As this is practically identical to the speed of light, he concluded that light must also be an electromagnetic phenomenon. Maxwell died at a young age in 1879, before his hypotheses could be confirmed experimentally.

Color Photograph

Maxwell was a real Scot. It was therefore no surprise that he chose tartan for the world's first color photograph. He gave his historical demonstration of color photography on May 17, 1861, to the Royal Institution of Great Britain in London.

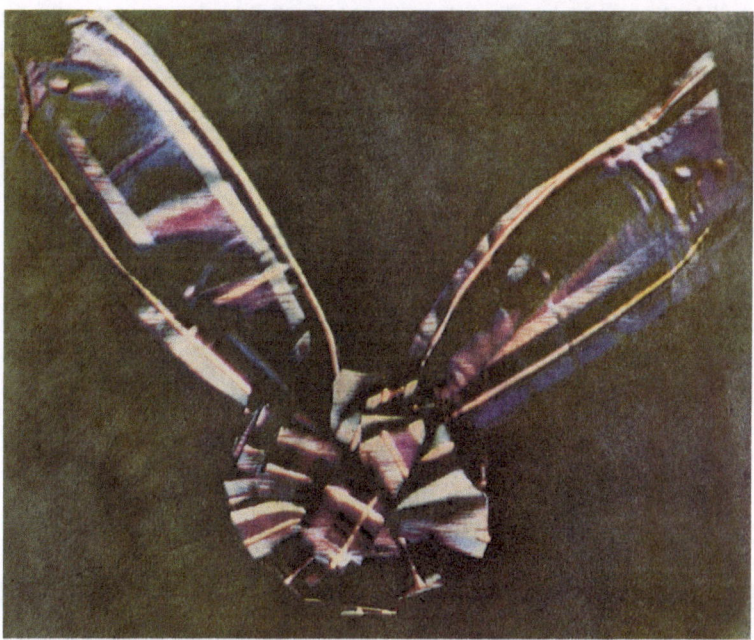

The first color photograph showed a cloth ribbon with a tartan pattern

Maxwell asked professional photographer Thomas Sutton to take three black-and-white photographs of a ribbon of tartan cloth, placing a red, a green and a blue filter in front of the lens. After Sutton had developed the photographic plates, he projected the three images on top of each other with red, green and blue light, so that together they produced a color reproduction of the ribbon. The image contained all the original colors of the tartan.

For a period of around 10 years, from his student years in Cambridge to his professorship in London, Maxwell observed and analyzed colors. Isaac Newton had made the first attempts at a quantitative color analysis in the early eighteenth century but, a 100 years later, it was largely Thomas Young who had taken a much greater step forward with his hypothesis that the eye has three color receptors, each receptive to a different part of the spectrum.

Maxwell elaborated on Young's three-color theory. Based on a model devised earlier by his former tutor, James Forbes, he made an ingenious disk for analyzing color. The disk consisted of two rings. The outer ring contained red, blue, and green segments, the size of which he could adjust himself. The inner ring comprised a black and a white segment, also adjustable in size. When the disk was rotated quickly, a gray color would appear in both rings. In the outer ring, the color depended on the surface area of the three primary colors. Similarly, the color in the inner ring depended on the ratio of the black to the white area. By experimenting with the areas of the segments, he was able to match the gray in the inner and outer rings.

The disk allowed Maxwell to describe the similarities between the three colors and the ratio between black and white. He described the values of each color in terms of the number of degrees covered by the segment concerned, for example: 0.37 red + 0.27 blue + 0.36 green = 0.28 white + 0.72 black. He then replaced one of

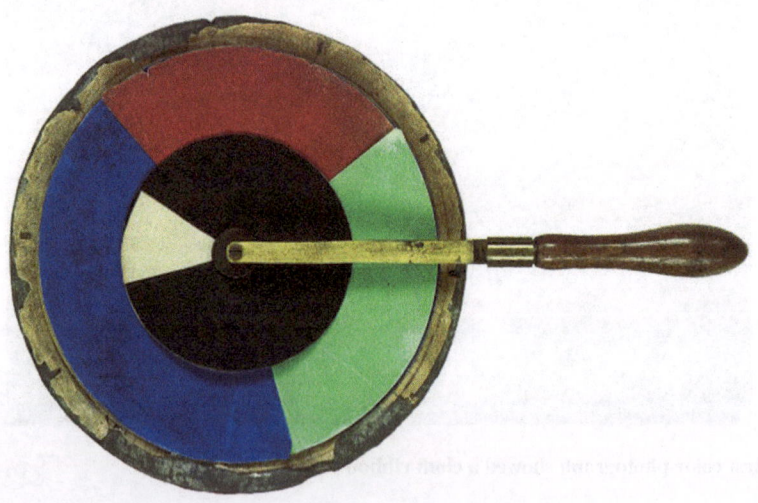

Maxwell's color disk

the three primary colors with a test color, for example, pale chrome. After repeating the procedure, he got 0.33 pale chrome + 0.55 blue + 0.12 green = 0.37 white + 0.63 black. Because the sum of the three primary colors and the sum of black and white is always 1, with a little arithmetic, every color can be described as a function of the three primary colors.

To make even more precise measurements, Maxwell designed a new instrument. His color box contained a number of prisms, which he used to separate sunlight into light of different wavelengths. He used this to draw up graphs describing the whole visible spectrum as a function of the three primary colors. These color functions, based on observations by himself and his wife Katherine, were the predecessors of those used by the International Commission on Illumination (CIE) in 1931 to define the formal color space. The CIE color space is still the standard for defining colors.

Maxwell had already conceived the theory behind the experiment with the color photograph in 1855, but the demonstration did not follow until 6 years later. Perhaps the most striking aspect of the experiment is that it cannot have worked entirely as intended, as the emulsions used at that time were not sensitive to the whole spectrum.

Exactly a 100 years after Maxwell gave his demonstration in London, the experiment was replicated by Ralph Evans and his colleagues at Eastman Kodak. They reconstructed it as closely as possible on the basis of old records by Maxwell and Sutton. The records show that the tartan cloth was photographed in clear sunlight against a background of black silk. The emulsion they used contained silver iodide as the light-sensitive material. Silver iodide is, however, only sensitive to wavelengths shorter than 430 nm, i.e., the extreme blue part of the spectrum. Without being aware of it, Maxwell and Sutton worked with photographic plates that were insensitive to much of the green part, and the entire yellow, orange, and red parts of the spectrum.

As a color filter, Sutton used glasses filled with colored solutions of metal salts: ammoniated cupric sulfate (blue), cupric chloride (green) and ferric thiocyanate (red). The negatives were printed on glass, using tannin, to produce a positive black and white image. The three plates were then projected on top of one another with the help of red, blue and green light from magic lanterns. As they had expected, Evans and his team obtained the same results as Maxwell and Sutton: the blue colors were very clear, and the green and red much less so. Maxwell's own record from that day shows that he, too, was not completely satisfied with the red and green part of the photograph but, in the end, the results were good enough to satisfy him and his audience.

Since the green spectrum lies approximately between 400 and 650 nm, a very small part of the green area lies below 430 nm. That concurs with Sutton's description, which says that the green photograph required an exposure time 120 times longer than the blue one. The wavelengths in the red part of the spectrum are higher than the green part, making it practically impossible for red light to be recorded on the silver iodide plates.

However, many red substances reflect not only red, but also invisible ultraviolet, which has a shorter wavelength than 400 nm. Unintentionally, Maxwell recorded ultraviolet on the sensitive glass plate which appeared red when he exposed it to red light. It was therefore pure luck that the photograph showed red at all. The experiment could actually only have been conducted completely correctly 15 years later, when the right emulsions had become available. Quite innocently, Maxwell and Sutton actually gave the first demonstration of false color photography.

Despite this defect, Maxwell's photograph was seen as the predecessor of today's color photographs, which are still based on the same principles. After his historic demonstration, it took several decades for the method to be applied in photography on a large scale. In 1907, brothers Auguste en Louis Lumière marketed the autochrome glass plate. The plates, coated with red, green, and blue starch grains, made it possible to produce a color photograph in a single exposure.

References

James Clerk Maxwell, 1860. 'On the Theory of Compound Colours and the Relations of the Colours of the Spectrum'. *Philosophical Transactions of the Royal Society of London* 150, 57–84.

Ralph Evans, 1961. 'Maxwell's Color Photograph'. *Scientific American* 205 (5), 118–126.

Richard Dougal et al., 2006. 'Then and now: James Clerk Maxwell and colour'. *Optics & Laser Technology* 38, 210–218.

Alexander Graham Bell

Alexander Graham Bell took only a
few years to invent the telephone. He
spent the rest of his creative life on
completely different experiments and
inventions, many for the benefit of
medical science. With his mechanical
breathing apparatus, a predecessor of
the iron lung, he hoped to save the
lives of drowning victims and premature babies.

R. Schils, *How James Watt Invented the Copier: Forgotten Inventions
of Our Great Scientists*, DOI 10.1007/978-1-4614-0860-4_15,
© Springer Science+Business Media, LLC 2012

Telephone

Alexander Graham Bell came from a family of elocution teachers. His father had devised "Visible Speech," a phonetic notation system with which deaf people and those with hearing difficulties could learn to speak. From 1868, Bell used the system to teach the hard of hearing, first in London and some years later in Boston. In 1873, he was appointed Professor of Vocal Physiology at Boston University, where he continued his quest for teaching aids for the deaf. Among other things, he investigated the transfer of sound using electrical signals. Bell's telephone was not the result of a linear, preconceived plan but of a variety of ideas that he was working on at the same time.

He used, for example, a human ear to build a phonograph to enable him to study how sound is converted into mechanical movement. By attaching a sort of pen to the ear bones, he was able to reproduce sound as visible waves.

He also devoted his attention to the electrical telegraph, the main form of communication at the time. The telegraph was only suitable for transmitting simple messages using short or long electrical pulses. Bell tried to adapt it so that multiple signals could be transmitted simultaneously. He thought that it must be possible to transmit a signal through continuous waves, instead of the discontinuous signals normally used in telegraphy. In 1875, with financial support from the parents of some of his deaf students, he succeeded in obtaining a number of patents on components of a "harmonic telegraph."

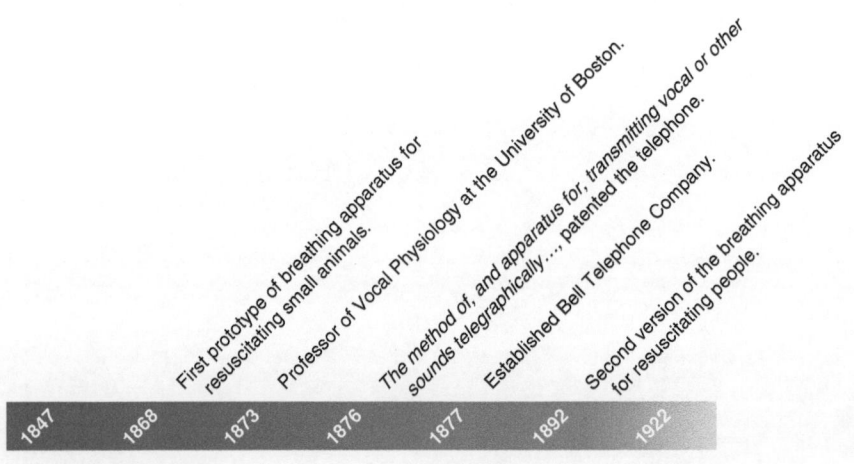

Alexander Graham Bell

On June 2, 1875, Bell and his assistant Thomas Watson conducted a crucial experiment, which made it clear to them that they must be able to convert variations in tone into a variable electrical signal. After this breakthrough, Bell began to describe the specifications of his telephone, which resulted in an approved patent on March 7, 1876. Bell's telephone converted voices into vibrations on a membrane, to which a permanent magnet was attached. The moving magnet generated an induction current in a coil so that the sound vibrations were converted into current variations. At the other end of the line, a similar device reversed the process, converting the current back into sound. Within a year, the invention was applied commercially and the Bell Telephone Company set up.

Bell's patent was one of the most controversial ever and led to a series of lawsuits. Bell was indeed not the only or the first to develop a telephone. In 1854, the Italian inventor Antonio Meucci had demonstrated his "telectrophone," but did not have enough money to apply for a patent. His lawsuit against Bell ended when Meucci died in 1889. His honor was somewhat restored posthumously when the American House of Representatives declared that, if Meucci had been able to submit his patent application, it would never have been issued to Bell.

Bell's most fascinating battle, however, was his conflict with Elisha Gray. On the very same day that Bell submitted his patent application, Gray turned up at the patent office with a claim for a similar design. He later accused Bell of altering his original patent after having seen Gray's application. A member of the patent office staff admitted 10 years later that he had indeed shown Bell Gray's application. But it was to no avail – Bell eventually came out on top in this case, too.

Artificial Respiration

From an early age, Alexander Graham Bell wanted to follow in his father's footsteps. In 1868, as part of his study in speech therapy, Bell took lessons in anatomy and physiology at the University of London. He became very interested in the physiology of breathing and the causes of oxygen deficiency. Bell was particularly concerned to find ways to save premature babies or other children weakened by respiratory problems. At that time, respiratory disorders among new-born babies and other young children were believed to be caused by their respiratory system not being fully developed or an insufficiently active central nervous system. It was assumed that children's lives could be saved simply by stimulating their breathing mechanically.

Bell reasoned that, if the muscles were too weak to move the thorax up and down, it should be quite simple to achieve the same effect by using an external force. He devised a rigid, airtight "vacuum jacket," which encased the entire upper body. A pump alternately raised and lowered the air pressure, automatically causing the patient to breathe in an out. Bell claimed that his "infant lifesaver" was much more effective than the methods used until then of mouth-to-mouth resuscitation and applying pressure to the upper body with the hands. According to Bell, his design was so efficient that it could, as it were, make a dead body breathe.

Bell's first prototype was large enough to encase the upper body of a cat. His initial experiments on a drowned cat showed that the principle indeed worked. The cat's chest moved up and down so that air flowed in and out of its lungs, but it did not survive the test.

In 1870, Bell emigrated to Canada with his parents and left the breathing apparatus behind in London. A few years later, he became Professor of Vocal Physiology at Boston and, with Thomas Watson, developed the first telephone. The telephone was a great commercial success and made him financially independent at a young age. That enabled him to pursue his own objectives and devote himself entirely to what he thought was important. The downside of his independence was that he worked too much in isolation. Partly for that reason, a lot of his work was forgotten, recorded only in his personal diaries and letters.

Bell did not return to his breathing apparatus until more than 10 years later. On August 25, 1882, he gave a presentation to the American Association for the Advancement of Sciences on a method of tracing metal in the human body. Completely unexpectedly, he ended the presentation with a description of his infant lifesaver.

Bell never explained his renewed interest in his invention, but it may have had something to do with his baby son, who had been born prematurely a year previously and died 3 days after the birth. Bell's wife had written in her diary: "He was a

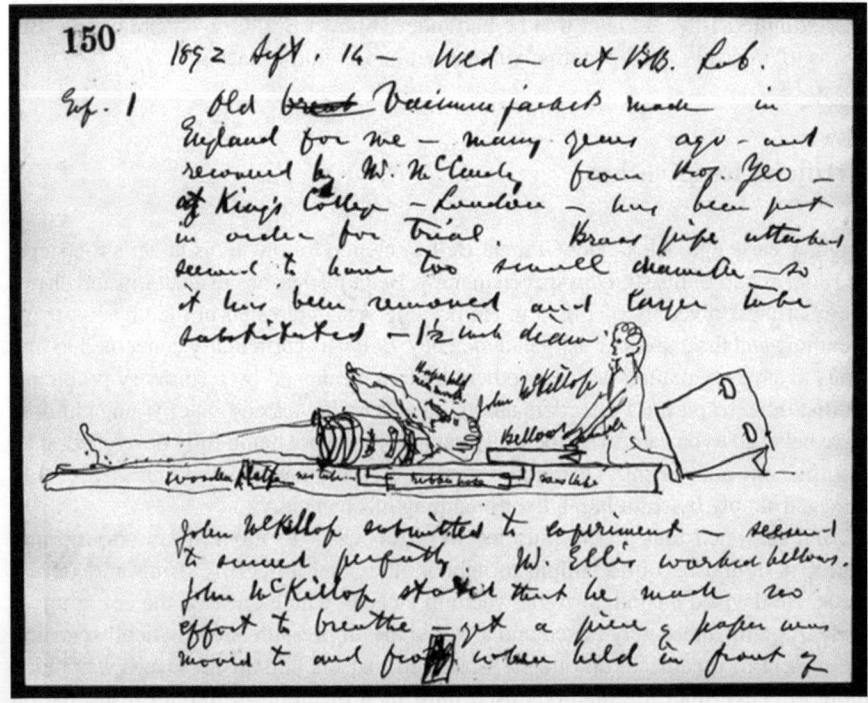

Original sketch of a vacuum jacket from Alexander Graham Bell's diary, September 14, 1892

strong little fellow and might have pulled through if they could only have established regular breathing."

In the years that followed, Bell's diary contained occasional sketches of new
designs for the breathing apparatus. In some of the drawings, the vacuum jacket is
replaced by a vacuum cabinet, in which a patient with weak breathing could sit. This
fluctuating interest in artificial respiration is typical of Bell's work. He would often
have an idea, invest all his energy into it, and then lose interest. Bell once observed
in one of his lectures that "It is often more interesting to observe the first totterings
of a child than the firm tread of a full-grown man."

In 1892, more than 20 years after leaving London, Bell wrote to Arthur McCurdy,
an old friend: "Don't forget when you go to London to try to find the apparatus I left
there 10–15 years ago for the production of artificial respiration. It was left at the
Alexandria Hotel, and Professor George Minchin, cousin of Dr. Chichester Bell,
undertook to have experiments with it at the University College, London. Make
every effort to find it and bring it to me." McCurdy found the machine and took it to
Bell. There was no evidence that it had been touched in the intervening 20 years.

In that same year, Bell built a version of the machine for human use. He was suddenly in such a hurry to complete the machine that he did not first take the time to
order the required components. Consequently, Mrs. Bell was surprised to discover
one morning that he had removed the bellows from her organ.

Alexander Graham Bell (extreme right) tests his breathing apparatus on a drowned sheep

Bell first tried the machine on a patient who could not breathe self-sufficiently for short periods. The breathing apparatus raised and lowered the patient's chest as intended, and a piece of paper moving in front of the mouth convinced Bell that the principle worked. He then had a sheep drowned and used his invention to bring it back to life. One of his staff members was so shocked by these devilish practices that he refused to accept his pay.

As with many of Bell's other inventions, his breathing apparatus did not attract much attention in the scientific world. However, the principle of respiration by external pressure – also known as negative pressure respiration – was applied on a large scale in the 1940s in the "iron lung." Inoculation against polio was not yet widespread, and the disease was still common. If it affected the respiratory muscles, patients would be placed in an iron lung for several weeks or even months. They would lie completely enclosed, except for the head, in a large cabin, in which fluctuating pressure would regulate their breathing. It was later discovered that positive pressure respiration – actively blowing air into the lungs – produced better results and negative pressure respiration is now rarely used.

Bell's interest in respiration was aroused once more in 1905 by an article in the Scientific American. The paper reported on the invention of a breathing apparatus in Hungary, which closely resembled Bell's machine, but without referring to it. He immediately wrote to the editors to inform them of his presentation in 1882, to prove beyond doubt that he had the idea first.

References

Alfred Henderson, 1972. 'Resuscitation Experiments and Breathing Apparatus of Alexander Graham Bell'. *Chest* 62 (3), 311–316.
Robert Bruce et al., 1997. *Alexander Graham Bell: The Life and Times of the Man Who Invented the Telephone*. Harry N. Abrams, 304 pp.
Thomas Baskett, 2005. 'Alexander Graham Bell and the vacuum jacket for assisted respiration'. *Resuscitation* 63, 115–117.

Hendrik Antoon Lorentz

The Netherlands' greatest physicist, Hendrik Antoon Lorentz, was awarded the Nobel Prize for his electron theory. As chairman of the State Commission for the Zuiderzee, Lorentz was responsible for investigating the effects of building a giant dam to seal off the Zuiderzee from the North Sea.

R. Schils, *How James Watt Invented the Copier: Forgotten Inventions of Our Great Scientists*, DOI 10.1007/978-1-4614-0860-4_16,
© Springer Science+Business Media, LLC 2012

Electron Theory

Hendrik Antoon Lorentz was one of the last great figures in classical physics, but his work was of great significance for the development of modern physics. He is seen, with good reason, as the spiritual link between James Clerk Maxwell and Albert Einstein.

Lorentz studied mathematics and physics at the University of Leiden where, at the age of 22, he successfully defended his thesis, *Over de theorie der terugkaatsing en breking van het licht* (On the theory of reflection and refraction of light). Several years later, he was appointed the first Dutch professor in theoretical physics. He continued Maxwell's study of the relationship between electricity, magnetism, and light. In 1865, Maxwell showed that light is actually nothing more than an electromagnetic wave phenomenon. At the time, however, scientists had great difficulty in interpreting the space in which light moved. As they could not imagine that light could move through a vacuum, they believed that there must be some kind of medium, in the same way that water is the medium in which water waves move. They proposed a sort of elastic substance, which they called the "ether." It proved impossible, however, to explain the observed properties of light in terms of prevailing views, in which Newton's classical mechanics played a central role.

Lorentz himself was also convinced of the existence of the ether, but did not adhere to an underlying mechanical model. He claimed rather that the ether is completely immobile, so that electromagnetic waves cannot drag it along with them. In 1892, he published a theory on the interaction between the ether and moving charged particles. In the first instance, he called these particles "ions," but Johnstone Stoney later introduced the term "electrons." The force that moving

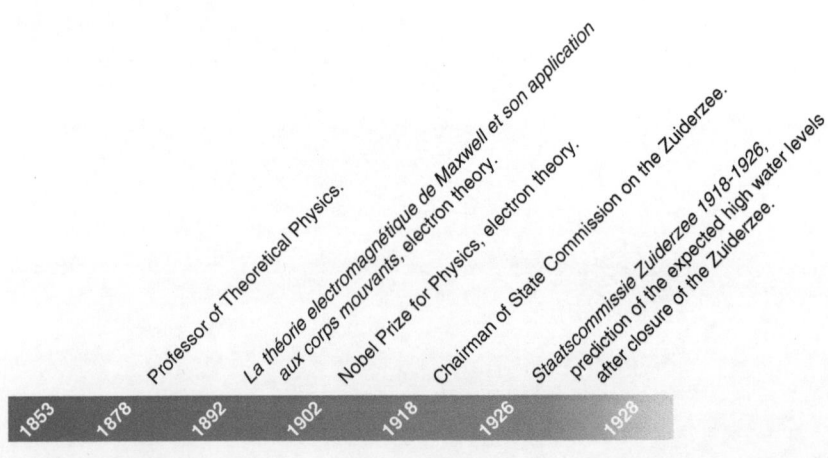

Hendrik Antoon Lorentz

charged particles experience in an electromagnetic field is known today as the "Lorentz force."

In a retrospective look at Lorentz' life, Einstein wrote that many physicists were not aware of just how groundbreaking his ideas were. One of these was the Lorentz–Fitzgerald contraction which, according to Lorentz, was required to explain the results of the Michelson–Morley experiments. These experiments were conducted to measure the speed of light through the ether, by comparing the speed of light from different directions. The basic assumption was that light moving in the same direction as the Earth would have a different speed than light moving perpendicular to it. The results showed, however, that the speed of light is the same in all directions.

To explain this apparent contradiction, Lorentz proposed that moving bodies that approach the speed of light contract in the direction of the movement. This was later known as the Lorentz contraction. It was Einstein, however, who stated that the speed of light was a universal constant which is the same for all observers, no matter in what direction or what speed they are moving. The Lorentz contraction is thus a logical consequence of the relative speed of different observers.

Enclosure Dam

In 1918, the Dutch government set up a commission to calculate the impact of closing off the Zuiderzee, the large inland sea in the north of the Netherlands, on coastal sea-water levels during gale conditions. In the bigger picture, the expected high water levels actually played a secondary role in the debate on whether to close off and reclaim the Zuiderzee. However, during the parliamentary debate on the legislation to approve the plans, Minister for Water Management Cornelis Lely had to promise that he would set up a commission to investigate the effects. He appointed prominent Nobel Prizewinner Hendrik Antoon Lorentz as chairman of the commission.

The inhabitants of the areas to the north of the planned "Enclosure Dam" not only feared for their safety, but also expected the plans to have a disadvantageous effect on their farmland. Lambertus Mansholt, a farmer and councilor from Groningen, summed up the disadvantages in a booklet entitled *De afsluiting der Zuiderzee, een ernstig gevaar voor Friesland en Groningen* (Closing off the Zuiderzee: A serious threat to Friesland and Groningen). According to Mansholt, without extra security measures, closing off the Zuiderzee would be a disaster for the north of the country.

The first plans to close off the Zuiderzee dated back to the seventeenth century. Hendric Stevin presented an ambitious plan that included the Zuiderzee and large parts of the Waddenzee. Since then, a number of proposals had been considered. But none had ever been put into practice, either because they were technically unfeasible, or because of a lack of funds or political will. Eventually, in 1886, a successful initiative was born, with the establishment of the Zuiderzee Association. Under the leadership of engineer Cornelis Lely, the Association explored the possibilities for reclaiming the Zuiderzee. Six years later, Lely's plans were ready, but the political decision-making process did not really get off the ground until after the disastrous floods of 1916. During a storm which raged for several days, the Zuiderzee flooded parts of the adjoining provinces of North Holland and Friesland. Even the Gelderse Vallei region further inland, was inundated. To spare the country from the threat of the water in the future, it was decided that the Zuiderzee should be closed off. That would shorten the Dutch coastline by 250 km in one fell swoop. And the construction of the Enclosure Dam would make the reclamation of land for farming and housing easier.

The State Commission for the Zuiderzee, also known as the Lorentz State Commission or the Storm Tide Commission, was given the following mandate: "To determine to what extent, as a consequence of the closure of the Zuiderzee, higher water levels and increased wave activity than is now the case can be expected along the coast of the mainland of North Holland, Friesland and Groningen, and of the North Sea islands lying of that coast."

Lorentz was not entirely new to the study of fluids. He had earlier written a paper on fluid dynamics. Perhaps this interest arose from his work on the ether and his hypothesis that it moved in the same way as an incompressible fluid. Nevertheless, it was a relatively new area of work for him, and his own words show that he did not take his new assignment lightly: "When the government asked us in 1918 to study the effects of the closure of the Zuiderzee on water levels during storm tides, I was personally alarmed. I must honestly admit that I felt a little intimidated. Physicists are not accustomed to tackling problems of such complexity with such a lack of reliable data." Fortunately, Lorentz did not have to face the challenge alone. He had the support of a substantial team of engineers, oceanographers and meteorologists.

The geography of the Zuiderzee and Waddenzee *before* and *after* closure. The numbers show the channel system according to the most complex model. The theoretical network is shown on the *right*

Everyone agreed that closing off the Zuiderzee would result in higher coastal water levels. The crucial question was, how much higher?

The Zuiderzee is connected to the North Sea by a number of narrow channels between the Frisian Islands. During a storm, the area to the south of the islands therefore fills up relatively slowly. If the Zuiderzee were to be closed off, that area would be reduced by a third, meaning it would fill up more quickly and water levels would be higher. Estimates varied from 15 to more than 35 cm. Because the dykes around the Waddenzee would have to be raised proportionately, it was important in terms of costs, if for no other reason, to make a more accurate estimate.

In general terms, Lorentz's approach was to build a model replicating the currents and water levels of the existing situation, before closure of the Zuiderzee. Then they would compare the model's prediction with historical data. If the model was accurate, it could then be applied to the new situation, with the Enclosure Dam in place.

The preceding debate had produced three possible methods of predicting high water levels after the Zuiderzee had been closed off. Two came from existing techniques in oceanography, but Lorentz chose a third alternative based on mathematical equations that describe the motion of fluids. In Lorentz's model, the Zuiderzee and Waddenzee were portrayed as a network of channels through which the tidal currents could flow into the area. Because the topography of the seabed was very

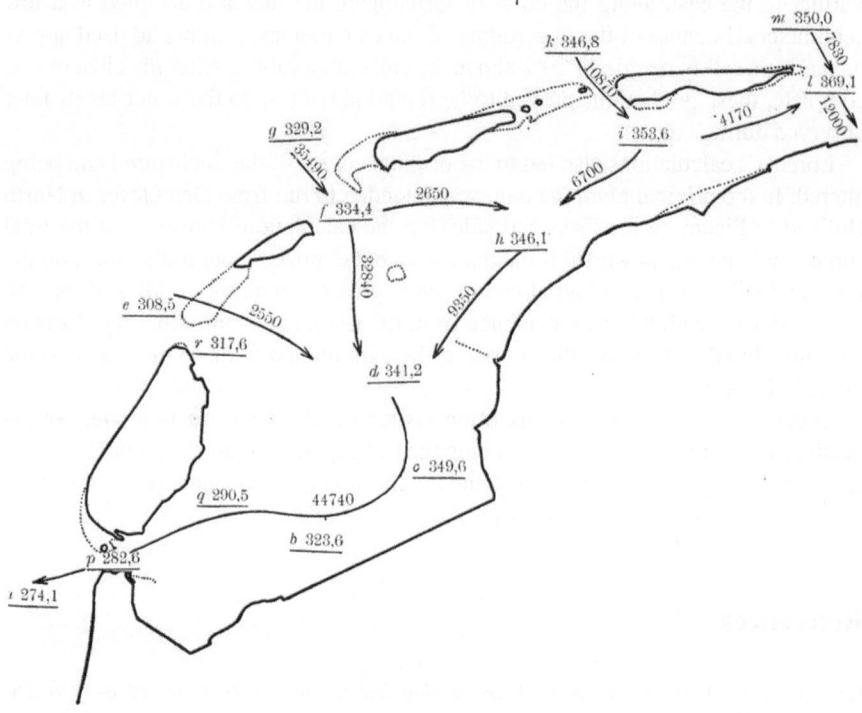

Current speeds (m³/s) and high water levels (cm) calculated by Lorentz for the storm of December 22 and 23, 1894, using the channel network method

variable, with considerable differences in depth over short distances, it was impossible to make a multidimensional model. Lorentz modeled the lateral depth variations in channels in the form of parallel channels with different depths. Using a complex iterative mathematical process, he was able to calculate the parameters of the whole system of channels, using historical measurements of the water levels during 50 earlier storms.

Before the method was applied to the complex situation around the Zuiderzee, it was first tested on two relatively simple water systems, the Gulf of Suez and the Bristol Channel. In both cases, the predicted changes in wave behavior and water levels concurred closely with historical data.

After these encouraging results, the method was applied to the existing situation in the Zuiderzee and Waddenzee, from the channels between the Frisian Islands to the coastline of the Zuiderzee. Lorentz used both a simple model with 17 channels and a more complex one, with 120 channels and subchannels. The calculations required were very labor intensive and were performed, mainly by Johannes Theodoor Thijsse, using slide rules and primitive mechanical calculators. After the existing situation had been successfully modeled, the calculations were repeated on the basis of a hypothetical closure of the Zuiderzee. The commission concluded that the expected increases in water level during storm tides would vary from 130 cm at the village of Piaam in Friesland to 60 cm at Harlingen, a little further to the north. Further to the east, along the coast of Groningen, the increase declined to a few centimeters. Because of the uncertainty of the calculations, Lorentz advised applying an extra safety margin of 20% above the calculated values. After the closure was complete, these predictions proved to be remarkably close to the water levels later observed during storms.

Lorentz's calculations also led to the original course of the Enclosure Dam being altered. In the original plan, the dam was intended to run from Den Oever in North Holland to Piaam on the Friesland side, but the calculations showed that the tidal surge would be less powerful if the dam were to be moved more to the north on the Friesland side. As doubts had already been expressed about the solidity of the soil around Piaam, with this new evidence from the Lorentz Commission, the decision was quickly taken to move the course of the dam around 5 km to the north, to the village of Zurich.

Together with Thijsse, Lorentz often visited the dam construction site, but his death in 1928 prevented him from seeing the last gap in the dam closed in 1932. The Lorentz sluice gates in Kornwerderzand are, however, a permanent reminder of his contribution to this ambitious project.

References

Hendrik Antoon Lorentz (chairman), 1926. *Verslag Staatscommissie Zuiderzee 1918–1926*. The Hague, 345 pp.
Geertruida de Haas-Lorentz (ed.), 1957. *H.A. Lorentz. Impressions of his Life and Work*. Amsterdam, 157 pp.

H.K. Kuiken, 1996. 'H.A. Lorentz: Sketches of his work on slow viscous flow and some other areas in fluid mechanics and the background against which it arose'. *Journal of Engineering Mathematics* 30: 1–18.

Kees Vreugdenhil et al., 2001. 'Waterloopkunde; een eeuw wiskunde en werkelijkheid'. *Nieuw Archief voor Wiskunde* (NAW) 5/2 nr. 3: 266–276.

Svante Arrhenius

Swedish Nobel Prizewinner Svante Arrhenius discovered that dissolved salts split into positively and negatively charged particles. Without him being aware of it, a digression to study ice ages made him the father of climate change science. More than a century ago, Arrhenius calculated the heat absorption of atmospheric carbon dioxide and predicted that a doubling of CO_2 levels would lead to a temperature increase of 4–6°C.

R. Schils, *How James Watt Invented the Copier: Forgotten Inventions of Our Great Scientists*, DOI 10.1007/978-1-4614-0860-4_17,
© Springer Science+Business Media, LLC 2012

Ions

A thesis is usually only the first step in a scientist's career, but for Svante August Arrhenius it was a turning point in his life. His thesis contained the first indications that dissolved salts dissociate into electrically charged particles, known as ions. The phenomenon of electrolysis, the separation of dissolved salts by an electrical current, had already been discovered by Michael Faraday. But Arrhenius suspected that dissolved salts always divide into ions, even without the application of an electrical current. This was a whole new perspective on chemical reactions, but at the time his professors were not overly impressed. After defending his thesis for 4 h, Arrhenius was given the low grade *non sine laude approbatur*, "not without praise accepted." Arrhenius would never forget this injustice, especially as it prevented him from teaching chemistry at the university.

Dissatisfied with the result, he sent his findings to other chemists in Europe, including Wilhelm Ostwald and Jacobus Henricus van 't Hoff. They realized the importance of Arrhenius' work and arranged a scholarship which enabled him to travel around Europe for several years, visiting the most prominent laboratories of the time. In this inspiring environment, he had the opportunity to develop his theory further.

The principle of electrolysis formed a kind of dividing line between physicists and chemists. Physicists generally ignored the fact that electrolysis involved the dissolution of salts, focusing on the transport of ions through the solution from one

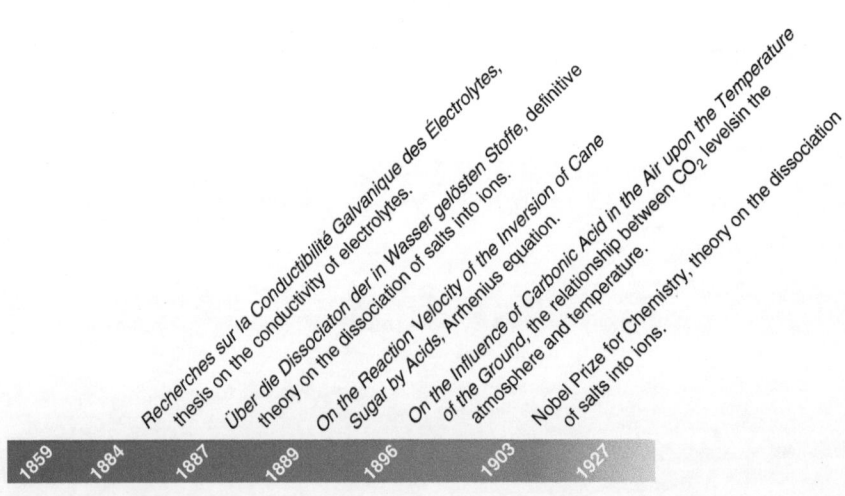

Svante Arrhenius

electrode to another. This is where the names of the ions originate from: cations move toward the cathode, and anions to the anode. The terminology was intended only to indicate on which electrode the electrolytic deposit would form, and said nothing about the properties of the ion itself.

The chemists, on the other hand, concentrated on the electrolytes, the dissolved salts, and the deposits on the electrodes. They devoted particular attention to the reaction between the solution, mostly water, and the electrolyte, whereby they assumed that dissolved salts were complex molecular aggregates.

In 1887, Arrhenius published his fully elaborated theory on the dissociation of salts in ions. He showed that dissolved salts always separate into ions, even if a current is not passed through the solution. Reactions in solutions always occur between the ions and not, as had been assumed until then, between the unseparated salts.

The ion theory was not accepted without criticism. August Friedrich Horstmann, a theoretical chemist, called the new movement *"Das wilde Heer der Ionier"* (The wild Horde of the Ionists), a reference to a sort of hunting party of supernatural origin. It left no doubt about the intentions of the "ionists."

Two years later, Arrhenius developed the concept of "activation energy," the energy required for a chemical reaction to occur. The Arrhenius equation describes the relationship between activation energy, temperature, and the rate of the reaction.

Arrhenius gradually became a respected scientist and, in 1891, he was finally granted the professorship at Stockholm University that he had so long coveted.

Greenhouse Effect

At the end of the nineteenth century, the Physics Society in Stockholm held a series of lectures and discussions on the cyclic pattern of cold and warm periods in the Earth's climate. Svante Arrhenius and his fellow researchers were looking for the causes of ice ages. In Arrhenius' time, understanding of recent ice ages – in the past 2 million years – was comparable to our current knowledge of climate change in the Pre-Cambrian period, more than 500 million years ago. The possible explanations presented in the lectures came from a variety of scientific disciplines, including astronomy, geography, geodesy, and physics.

It was the presentation by geologist Arvid Högbom that put Arrhenius on the trail of atmospheric carbon dioxide (CO_2). Högbom described the most important processes in the global carbon cycle. He was especially interested in long-term changes in the exchange of carbon between the atmosphere and the lithosphere, as witnessed in the weathering of rocks. But it did not escape Högbom that burning coal would lead to an increase in the level of CO_2 in the atmosphere. He calculated that combustion of the 500 million tons of coal produced at that time per year would increase the CO_2 level in the atmosphere by only a thousandth part. According to Högbom, this was roughly equal to the annual amount of CO_2 sequestered through the formation of calciferous rocks.

Högbom was also aware that the oceans acted as a buffer. He stated that a sudden doubling of the level of CO_2 in the atmosphere, for example, after an enormous volcanic eruption, would not ultimately lead to twice as much atmospheric CO_2 because the oceans would absorb more than usual. Högbom's analysis of the global carbon cycle was very remarkable, but it was Arrhenius who made the link between variations in atmospheric CO_2 levels and variations in the climate.

Arrhenius started to work this idea out in greater detail in 1894, on the basis of earlier work by Joseph Fourier and John Tyndall. In 1824, Fourier had published *Remarques générales sur les températures du globe terrestre et des espaces planétaires* (General remarks on the temperature of the earth and outer space), in which he presented the idea of the Earth's atmosphere as a sort of greenhouse. He referred to the presence of gases in the atmosphere that allow sunlight to pass through, but which absorb the heat radiated from the Earth's surface.

Irishman John Tyndall was the first to determine the absorption of radiation by gases experimentally, in 1859. With a self-designed spectrophotometer, fitted with a tube in which he could place gases under pressure, he identified in the laboratory the absorption for water vapor, carbon dioxide, methane, nitrous oxide, ozone and various organic molecules. Tyndall focused his attention mainly on the importance of water vapor for the Earth's heat balance: "Aqueous vapour is a blanket, more necessary to the vegetable life of England than clothing is to man. Remove for a single summer night the aqueous vapour from the air which overspreads this

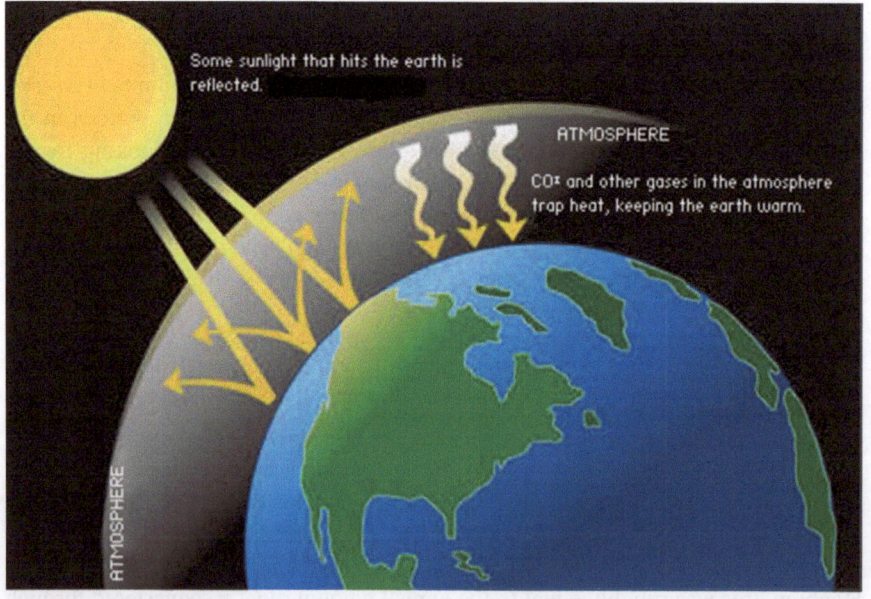

The greenhouse effect

Latitude	Carbonic Acid=0.67					Carbonic Acid=1.5					Carbonic Acid=2.0					Carbonic Acid=2.5					Carbonic Acid=3.0				
	Dec.-Feb.	March-May	June-Aug.	Sept.-Nov.	Mean of the year	Dec.-Feb.	March-May	June-Aug.	Sept.-Nov.	Mean of the year	Dec.-Feb.	March-May	June-Aug.	Sept.-Nov.	Mean of the year	Dec.-Feb.	March-May	June-Aug.	Sept.-Nov.	Mean of the year	Dec.-Feb.	March-May	June-Aug.	Sept.-Nov.	Mean of the year
70	-2.9	-3.0	-3.4	-3.1	-3.1	3.3	3.4	3.8	3.6	3.52	6.0	6.1	6.0	6.1	6.05	7.9	8.0	7.9	8.0	7.95	9.1	9.3	9.4	9.4	9.3
60	-3.0	-3.2	-3.4	-3.3	-3.22	3.4	3.7	3.6	3.8	3.62	6.1	6.1	5.8	6.1	6.02	8.0	8.0	7.6	7.9	7.87	9.3	9.5	8.9	9.5	9.3
50	-3.2	-3.3	-3.3	-3.4	-3.3	3.7	3.8	3.4	3.7	3.65	6.1	6.1	5.5	6.0	5.92	8.0	7.9	7.0	7.9	7.7	9.5	9.4	8.6	9.2	9.17
40	-3.4	-3.4	-3.2	-3.3	-3.32	3.7	3.6	3.3	3.5	3.52	6.0	5.8	5.4	5.6	5.7	7.9	7.6	6.9	7.3	7.42	9.3	9.0	8.2	8.8	8.82
30	-3.3	-3.2	-3.1	-3.1	-3.17	3.5	3.3	3.2	3.5	3.47	5.6	5.4	5.0	5.2	5.3	7.2	7.0	6.6	6.7	6.87	8.7	8.3	7.5	7.9	8.1
20	-3.1	-3.1	-3.0	-3.1	-3.07	3.5	3.2	3.1	3.2	3.25	5.2	5.0	4.9	5.0	5.02	6.7	6.6	6.3	6.6	6.52	7.9	7.5	7.2	7.5	7.52
10	-3.1	-3.0	-3.0	-3.0	-3.02	3.2	3.2	3.1	3.1	3.15	5.0	5.0	4.9	4.9	4.95	6.6	6.4	6.3	6.4	6.42	7.4	7.3	7.2	7.3	7.3
0	-3.0	-3.0	-3.1	-3.1	-3.02	3.1	3.1	3.2	3.2	3.15	4.9	4.9	5.0	5.0	4.95	6.4	6.4	6.6	6.6	6.5	7.3	7.3	7.4	7.4	7.35
-10	-3.1	-3.0	-3.2	-3.1	-3.12	3.2	3.2	3.2	3.2	3.2	5.0	5.0	5.2	5.1	5.07	6.6	6.6	6.7	6.7	6.65	7.4	7.5	8.0	7.6	7.62
-20	-3.1	-3.1	-3.3	-3.2	-3.2	3.2	3.2	3.4	3.3	3.27	5.2	5.3	5.5	5.4	5.35	6.7	6.8	7.0	7.0	6.87	7.9	8.1	8.6	8.3	8.22
-30	-3.1	-3.2	-3.4	-3.4	-3.35	3.4	3.5	3.7	3.5	3.52	5.5	5.6	5.8	5.6	5.62	7.0	7.2	7.7	7.4	7.32	8.6	8.7	9.1	8.8	8.8
-40	-3.3	-3.3	-3.3	-3.4	-3.37	3.6	3.7	3.8	3.7	3.7	5.8	6.0	6.0	6.0	5.95	7.7	7.9	7.9	7.9	7.85	9.1	9.2	9.4	9.3	9.25
-50	-3.4	-3.4	—	—	—	3.6	3.7	—	—	—	6.0	6.0	—	—	—	7.7	7.9	—	—	—	9.1	9.2	—	—	—
-60	-3.2	-3.3	—	—	—	3.6	3.7	—	—	—	6.0	6.1	—	—	—	7.9	8.0	—	—	—	9.4	9.5	—	—	—

The effect of changes in CO_2 levels on temperature, according to Arrhenius. The change in the CO_2 level (carbonic acid) varies from a factor of 0.67 to a factor of 3.0. The effect of CO_2 is calculated for the four seasons and with steps of 10° latitude (Latitude)

country, and you would assuredly destroy every plant capable of being destroyed by a freezing temperature. The warmth of our fields and gardens would pour itself unrequited into space, and the sun would rise upon an island held fast in the iron grip of frost."

Arrhenius' climate model took account of the heat absorption by carbon dioxide and water vapor. Because he did not have access to reliable direct measurements of absorption in the Earth's atmosphere, he used a series of measurements by American astronomer Samuel Langley. The Langley series, dating from 1890, consists of observations of the heat the Earth receives from the Moon, from which Arrhenius was able to deduce the absorption of carbon dioxide and water vapor.

Arrhenius then calculated the change in the temperature on Earth if the atmospheric CO_2 level were to change by a factor ranging from 0.67 to 3.0. He also calculated the effect in steps of 10° latitude, from 70° North to 60° South, and again for the four seasons. A tedious calculation, as he himself called it. But as a devotee of analyzing large datasets, he did not shrink from the task.

He came to the conclusion that a fall in the CO_2 level by 33% would lead to a temperature drop of 2.9–3.4°C. During the last ice age, the temperature was some 4–5°C lower, allowing Arrhenius to calculate that the CO_2 level must have then been around 40% lower.

We now know that lower CO_2 levels were not the main cause of the ice ages, but that they did play an important role in reinforcing the lower temperatures. Currently, the Milankovic theory provides the most convincing explanation of ice ages. Fluctuations in the position of the Earth's axis and periodic changes in its orbit cause small changes in the amount of sunlight reaching Earth. In addition, the position of the continents close to the poles, especially in the northern hemisphere, plays a significant role. These large, snow-covered expanses reflect the sunlight and intensify the cooling process.

The current generation of climate scientists do not, however, refer to the work of Arrhenius in connection with ice ages, but with the opposite phenomenon of global warming. Although Arrhenius was mainly interested in the "cold" side of the equation, his climate model also showed that a doubling of the CO_2 level would lead to a temperature rise of 4.0–6.1°C. Current insights suggest a temperature rise of 2–3°C, meaning that Arrhenius' prediction was inaccurate by a factor of two. That is not so surprising, given the limited knowledge and data at his disposal. Climate science is very complex and the global carbon cycle, with all its feedbacks, is still not fully understood today. With hindsight, the article by Langley, on which Arrhenius based his calculations, contained an incorrect extrapolation of measurement data. Ten years after his original publication, Langley himself produced a rectification, but Arrhenius never incorporated it into his calculations. Not surprisingly, as the search for the causes of the ice ages was for him no more than a digression.

Arrhenius was actually not overly concerned about our climate, as he estimated that it would take 3,000 years to increase CO_2 levels by half by burning coal. At that time, of course, he could have no inkling of the explosive development in the use of fossil fuels that would occur later. Besides, he saw climate warming more as a benefit than a threat. During a lecture, he once expressed the "pleasant thought" that our descendants, albeit after many generations, would live in milder climatic conditions. Not a strange idea for someone from a cold country like Sweden.

References

Svante Arrhenius, 1896. 'On the Influence of Carbonic Acid in the Air upon the Temperature of the Ground'. *Philosophical Magazine* 41, 237–276.

Elisabeth Crawford, 1996. *Arrhenius: From Ionic Theory to the Greenhouse Effect*. Science History Publications, 320 pp.

Martin Heimann, 1997. 'A review of the contemporary global carbon cycle and as seen a century ago by Arrhenius and Högbom'. *Ambio* 26 (1), 17–24.

Edouard Bard, 2004. 'Greenhouse effect and ice ages: historical perspective'. *Comptes Rendus Geoscience* 336, 603–638.

Pierre Curie

Together with his wife Marie, Pierre
Curie discovered the radioactive elements polonium and radium. Several
years earlier, working with his brother
Jacques, he had discovered the piezoelectric effect in crystals.

R. Schils, *How James Watt Invented the Copier: Forgotten Inventions of Our Great Scientists*, DOI 10.1007/978-1-4614-0860-4_18,
© Springer Science+Business Media, LLC 2012

Radioactivity

In the spring of 1894, Pierre Curie met his future wife Marie Sklodowska. After successfully completing her studies at the Sorbonne, Marie decided to write her dissertation under the supervision of Henri Becquerel, who had shortly before discovered that uranium salts emitted a kind of radiation. Marie Curie would continue researching this phenomenon, known as becquerel or uranium radiation.

At that time, Pierre Curie was a researcher and teacher at the *École Municipale de Physique et de Chimie Industrielles* (Municipal School of Industrial Physics and Chemistry). In 1897, he arranged temporary accommodation for his wife at the institute, in the form of a shed equipped as a laboratory where she could conduct her research.

Becquerel had proved the existence of the radiation using a photographic plate, but that was not accurate enough for the Curies. Pierre Curie developed a special electrometer with which Marie could measure the ionizing effect of radiation from uranium and thorium. She concluded that the radiation was an inherent property of the atom itself, rather than the effect of some kind of chemical reaction. She called this property "radioactivity."

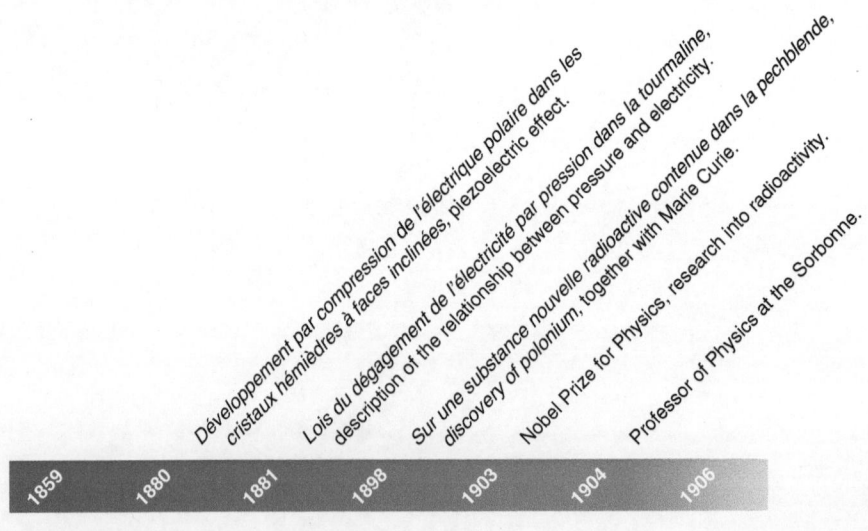

Pierre Curie

Because of their encouraging results, Pierre stopped his own research into crystals and magnetism completely, and he and Marie continued to study radioactivity. They focused their attention on the mineral uraninite, also known as pitchblende, in which they found higher radioactivity than in pure uranium. They suspected that pitchblende contained slight impurities that were responsible for the high radioactivity.

In 1898, in *Sur une substance nouvelle radioactive contenue dans la pechblende* (On a new, radioactive substance contained in pitchblende), they described the discovery of a new radioactive element. They called it polonium, a reference to Marie's homeland. Toward the end of that year, they succeeded in isolating a second element from pitchblende, which they called radium. Producing the polonium and radium was an enormous task. They used tons of pitchblende from a mine in Bohemia, from which – after 3 years – they isolated a tenth of a gram of radium chloride.

In 1903, Becquerel and the Curies were jointly awarded the Nobel Prize for Physics. In his acceptance speech, Pierre Curie warned that radium could be very dangerous if it fell into the wrong hands. He compared their discovery with that of dynamite by Alfred Nobel himself. Like Nobel, however, Curie believed that humanity would eventually reap more benefit than harm from such inventions.

Pierre Curie experimented with radium by exposing his own skin to the radiation. It rapidly became clear that it could cause astounding physiological reactions. Exposing his arm to radiation for several minutes resulted in a skin infection, comparable to the effects of X-ray radiation. In the early twentieth century, radium was therefore used to treat a wide variety of tumors. Around 1950, radium was gradually replaced by radioactive isotopes of cesium, cobalt, and iodine.

In 1904, Pierre Curie was appointed Professor of Physics at the Sorbonne, with Marie Curie in charge of his laboratory. Two years later, he was killed in a tragic accident and Marie took over his chair at the Sorbonne.

Piezoelectric Effect

In 1880, Jacques Curie announced to the French society of mineralogy that, together with his brother Pierre, he had discovered that exerting pressure on certain crystals generated an electrical charge at the extreme ends of the crystals. At that time, they could not possibly have imagined that, in the following century, their discovery, known as piezoelectricity, would be applied in almost all electronic devices.

The Curie brothers were both assistant researchers at the Sorbonne. Jacques assisted the mineralogist Charles Friedel and Pierre the crystallographer Paul Desains. Friedel was working on pyroelectricity, the generation of electrical currents by heating certain crystals. It had been known since the early eighteenth century that the crystal tourmaline attracted ash from glowing coals. This was first believed to be a kind of magnetism, but it was later linked to an electrical polarization of the crystal: one end of the crystal was positively charged and the other negatively.

It is not known exactly how the Curies discovered piezoelectricity, but they were probably inspired by the work of Friedel. In 1879, Friedel had conducted a series of experiments in which he heated the various faces of pyroelectric crystals. The pyro-electric effect proved to be dependent on the way in which the crystal was heated, but Friedel was unable to explain these differences.

The Curies thought that, if Friedel's observations were correct, there must be some form of distortion in the crystal structure. They concluded that the change of temperature caused a change in the structure, which in turn caused the electrical effect. If the distortion was generated in another way, for example by exerting pres-sure, an electrical effect should therefore also be observable.

To test their hypotheses, Pierre and Jacques cut the crystals into very thin slices, exactly along their axes. They then placed the crystal between two copper plates and an isolator, and used clamps to exert pressure on it. They measured the difference in charge between the two ends of the crystal using an electrometer designed by William Thomson that could detect very small variations in electrical potential.

They subjected dozens of different kinds of crystals to the pressure tests. They succeeded in demonstrating a piezoelectric effect in tourmaline, sphalerite, boracite, topaz, calamine, and quartz, but not in amorphous materials. They observed that the effect of exercising pressure on the piezoelectric crystals was comparable to that of lowering the temperature, while reducing the pressure produced the same effect as increasing the temperature.

After these comprehensive experiments, the Curies had sufficient data to deduce a number of general rules on the link between crystal structure and piezoelectricity.

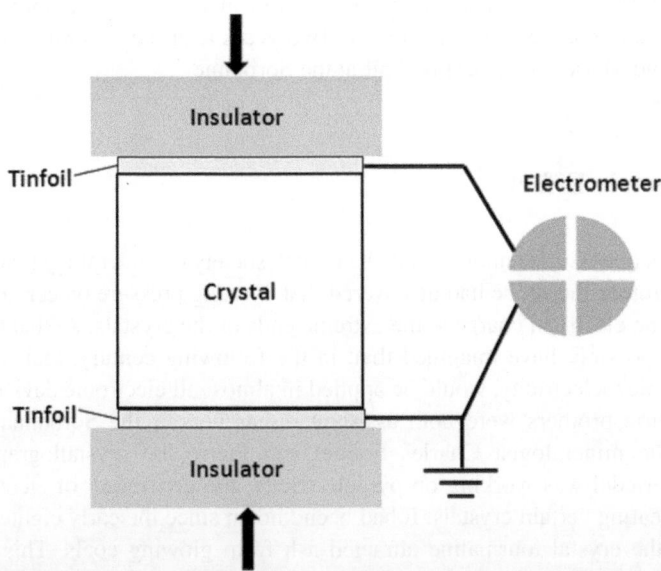

Diagram showing the Curie's equipment for measuring the piezoelectric effect. A piece of crystal was clamped between two isolators with a strip of tin foil on both sides to conduct the current to the electrometer. The *arrows* show where pressure was exerted

To generate an electric current between the two ends of the crystal's axis, the symmetry of the crystal had to comply with certain specific requirements. There must be no center of symmetry, no plane of symmetry perpendicular to the electrical axis, and no axis of symmetry of an even order perpendicular to the electrical axis. In addition to these rules of symmetry, they were able to deduce that the positive charge always occurs at the end of the axis at which the angle between the axis and the crystal's face is more acute.

In a second series of experiments, the Curies investigated the link between the pressure exerted and the electrical charge. They discovered that each kind of crystal has its own piezoelectric constant that describes the relationship between the pressure and the electrical potential. For tourmaline and quartz, they calculated a piezoelectric constant of 6.3×10^{-8} and 5.4×10^{-8} statcoulombs per dyne, respectively. These values vary less than 10% from those known today.

The first piezoelectric application that the brothers devised made clever use of the relationship between pressure and electrical potential. Their piezoelectrometer measures electrical potential from the pressure that has to be exerted on a crystal.

A year after the Curie brothers discovered piezoelectricity, Gabriel Lippmann stated, on the basis of thermodynamic calculations, that the converse piezoelectric effect must also exist. According to Lippmann, if an electrical current is applied to the ends of a crystal, there will be a shift in the crystal lattice. The Curies immediately took up the challenge, but found themselves facing a tough task. After all, a piece of crystal a centimeter thick would expand or contract by no more than a thousandth of a millimeter, a movement that could not be observed optically.

With their first attempt, they tried to design a nanometer, an instrument that should have been able to measure the pressure of the crystal. The nanometer did work, but not accurately enough to provide conclusive evidence.

They then devised an ingenious method that used a kind of lever to magnify the changes in the dimensions of the crystal by a factor of around 40. They were then able to read the magnified change through a microscope. This method was successful and they were able to prove the converse piezoelectric effect experimentally.

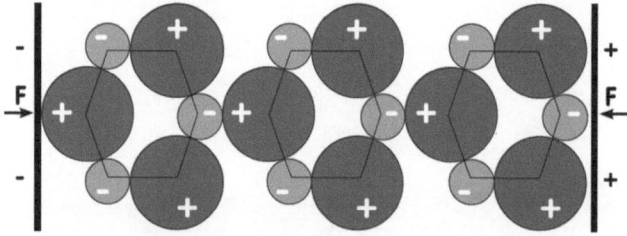

The principle of piezoelectricity in quartz (SiO_2). The silicone atoms (*large spheres*) have a positive charge and the oxygen atoms (*small spheres*) a negative charge. The force exerted (F) makes the oxygen atoms move a little to the left relative to the silicon atoms, resulting in a negative charge on the *left* and a positive charge on the *right*

It took more than 30 years for piezoelectricity to be applied outside the laboratory. The first practical application was developed during the First World War, in sonar systems to track German submarines. The system, which used the converse piezo-electric effect, was developed by Paul Langevin, one of Pierre Curie's students.

The piezoelectric element of the sonar consisted of thin quartz crystals glued between two small steel plates. When a very high frequency alternating current was applied, the crystal would expand and contract at the frequency of the current, producing an ultrasonic sound. Conversely, the same piezoelectric element could be used to receive ultrasonic signals and convert them into an alternating current signal.

Some years later, there was further progress, with the application of quartz crystals in oscillators. Quartz crystals cut in a specific way have a very stable unique frequency when subjected to an electric current. Quartz oscillators were first used to enable radio stations to broadcast at stable frequencies. Later, they appeared in all kinds of electronic apparatus requiring a stable frequency, including watches, computers, pacemakers, and mobile telephones. The Curies did not live to see any of these developments. When Jacques moved to Montpellier in 1883 to teach mineralogy, the brothers went their separate ways. Pierre stayed in Paris, where he and his wife later discovered the radioactive element polonium and radium. The famous photograph on the cover of the first issue of the magazine *Le Radium* shows Marie Curie working with the piezoelectrometer built by her husband and his brother.

References

Shaul Katzir, 2003. 'The Discovery of the Piezoelectric Effect'. *Archive for History of Exact Sciences* 57, 61–91.
Matthew Trainer, 2003. 'Kelvin and piezoelectricity'. *European Journal of Physics* 24, 535–542.
Philippe Molinié et al., 2006. 'Une application méconnue et pourtant célèbre de l'électrostatique: les travaux de Marie Curie, de la découverte du radium à la métrologie de la radioactivité'. *Journal of Electrostatics* 64, 461–470.

Walther Nernst

Walther Nernst is one of the founders of physical chemistry. His name is immortalized in the Nernst equation, and the third law of thermodynamics earned him the Nobel Prize for chemistry. Together with Siemens and Bechstein, he also developed an electric grand piano.

R. Schils, *How James Watt Invented the Copier: Forgotten Inventions of Our Great Scientists*, DOI 10.1007/978-1-4614-0860-4_19,
© Springer Science+Business Media, LLC 2012

Thermodynamics

Walther Nernst was educated at universities in Switzerland, Austria, and Germany. After completing his studies, he worked as an assistant to Wilhelm Ostwald. During that period, he met other pioneers in physical chemistry, including Svante Arrhenius and Jacobus Henricus van't Hoff. Together, they were known as the "ionists" because of their research into the role of ions in chemical reactions.

From 1887, Nernst devoted his attention to electrochemistry and, 2 years later, published his "Nernst equation," which can be used to calculate the potential of an electrode, and determine the voltage of batteries or other electrochemical cells.

In 1891, Nernst was appointed Professor of Chemistry at the University of Göttingen, where he set up a new physical chemistry institute some years later. Within a short time, Nernst and his staff had made the institute world famous.

As well as his interest in theoretical and fundamental science, Nernst had a good nose for practical and commercial applications. His greatest financial success was the Nernst lamp, later produced by the Allgemeine Elektrizitäts-Gesellschaft (AEG). The lamp contained a ceramic element that radiated a clear white light when heated. Unlike lamps with a carbon filament, the filament in the Nernst lamp also worked in

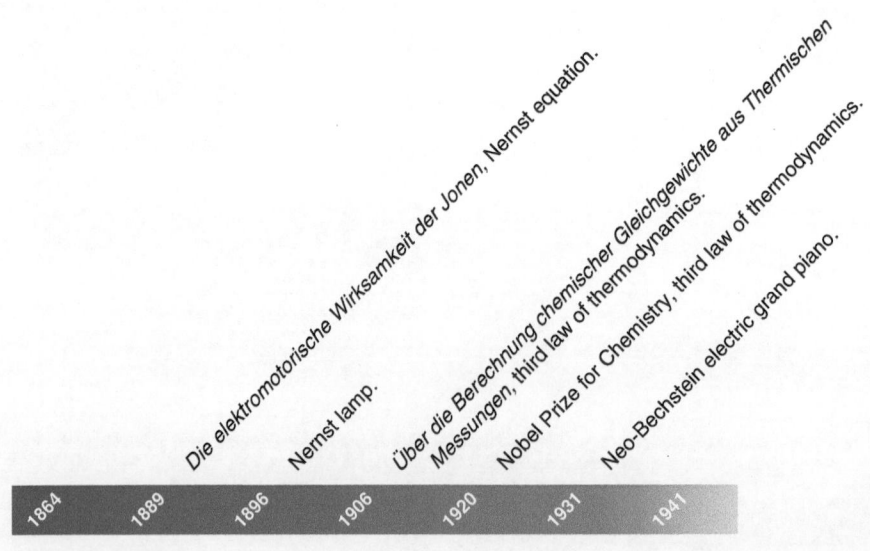

Walther Nernst

normal air. It did not therefore need to be placed in a vacuum or surrounded by incombustible gas. Moreover, the light was more natural and the lamp was twice as efficient as a lamp with a carbon filament. The greatest disadvantage of the Nernst lamp was that the ceramic element had to be preheated before it could conduct electricity. Around four million of the lamps were produced, but eventually Nernst's invention lost out to its competitors.

In 1905, Nernst moved to Berlin where he lived for the remainder of his active life. His first lectures in Berlin addressed the thermodynamics of chemical equilibrium reactions. Chemical equilibrium is the state in which the forward and reverse reactions are in equilibrium. Nernst sought a way to use thermodynamics to calculate the equilibrium position of chemical reactions.

The first and second laws of thermodynamics could not solve Nernst's problem. The first law refers to the conservation of energy, while the second states that entropy of a system – the degree of molecular disorder – increases until it is in equilibrium. In practical terms, the second law means, for example, that heat cannot flow from a colder to a warmer object. Although it was clear that entropy decreases as temperature falls, no one understood the exact relationship between changes in entropy and temperature.

During his lectures in Berlin, it occurred to Nernst that entropy is probably zero if the temperature reaches the absolute zero of −273.15°C. This breakthrough was not only of great theoretical importance, but also very valuable in practical terms. It was now possible to calculate the equilibrium constant, and therefore the position of a chemical equilibrium. This knowledge made industrial chemical equilibrium reactions, for instance as used in the production of nitrogen-based fertilizers, much more efficient.

Neo-Bechstein Grand Piano

The first musical instrument to make use of electricity was probably the *Clavecine Électrique*, built in 1759 by the Jesuit priest Jean-Baptiste de Laborde. The instrument used an electrostatic charge to move a clapper back and forth between two bells. The charge was stored in a "Leiden jar," an early form of capacitor.

It was not until the early twentieth century, however, that interest in electronic music really got off the ground. Until 1930, a variety of instruments had been developed in which electricity played a role of some kind in producing sound. Most of them had a keyboard, though the first prototypes of an electric guitar and violin had been demonstrated in 1927. Usually, the sound was produced mechanically and then amplified electrically, but there were also instruments with an electromagnetic sound source.

Although the musical establishment was not very enthusiastic, some innovative composers did write music especially for electronic instruments. This trend did not escape the notice of Walther Nernst. He had already shown with his Nernst lamp that he was capable of converting his scientific knowledge into successful commercial initiatives. He was convinced that he could apply his knowledge of physics to the further development of electronic instruments. His motives were not purely scientific; he expected, once again, that there was a tidy profit to be made.

Nernst, now a renowned scientist, succeeded in interesting two other world famous figures in the development of an electric grand piano. Bechstein, who had been building pianos since 1853 and was a leading manufacturer of concert pianos, was to build the instrument itself. The electrical components were to be supplied by Siemens & Haske, which had been producing components for telegraph networks since 1847 and, 80 years later, had become a large and diverse manufacturer of electronic components.

Eventually, Nernst and his partners built around 150 electric grand pianos, some of which still survive in museums. They were first marketed in 1931 as the "Neo-Bechstein," but were also known as the "Bechstein-Siemens-Nernst electric grand piano." The Neo-Bechstein was a modified acoustic grand piano. It was actually a semi-acoustic instrument, with the sound being generated in the traditional way by striking strings. The principle was simple: microphones captured the vibrations of the string and sent the signal to a loudspeaker through an electrical circuit.

The idea of using microphones to amplify the string sounds of a piano had already been proposed 3 years previously by Oskar Vierling, of the Heinrich Hertz Institute. They had picked up on the idea at the radio in Hamburg and tried to convert the vibrations of the strings directly into an alternating current signal that could then be directly broadcast. But the results were disappointing because, without the

Walther Nernst working on a one-string model of his electric piano

electrical signal being modified further, it proved impossible to replicate the complex timbre of a piano.

When a piano string is struck, it vibrates at various frequencies at the same time. Each string has its own fundamental frequency and a number of overtones. The timbre is determined by the mix of the fundamental frequency and the overtones. If, for example, the fundamental frequency is 440 Hz, the first three overtones will have frequencies of 880, 1320, and 1760 Hz, respectively. The fundamental frequency has the greatest volume, and the volume then decreases from the first to the highest overtone. The extent to which the sound is amplified and the overtones are audible depends partly on the resonance of the piano's sound board. In a piano without a sound board, like the Neo-Bechstein, the specific timbre and the amplification are created in the electrical circuit.

The Neo-Bechstein is a small grand piano, around a meter and a half shorter than a regular grand. The strings are shorter and thinner, and are struck by a specially developed micro-hammer, a small hammer fixed to the main hammer with a flexible leather band. The main hammer strikes against a small block, causing the small

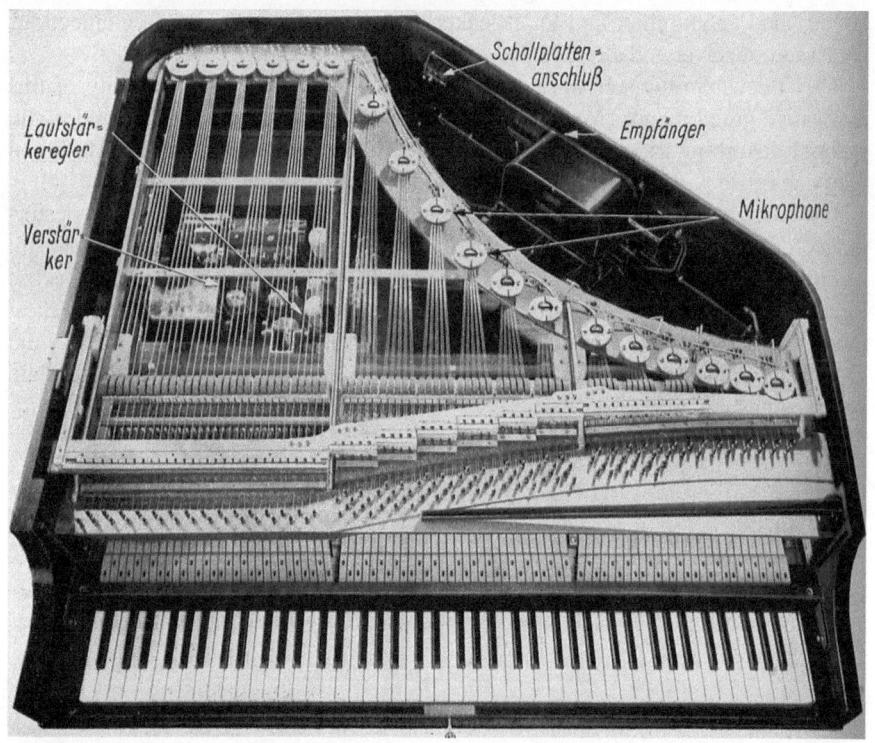

Interior of the Neo-Bechstein grand piano. To the *left* is the amplifier (Verstärker) and the volume regulator (Lautstärkeregler). *Right* is the connection for the record player (Schallplattenanschluss) and the radio (Empfänger). At the end of each group of five strings are the microphones (Mikrophone)

hammer to swing further and touch the string very lightly. The idea behind this was that the smaller the amplitude of the string, the purer the vibrations would be. Furthermore, the string does not have to be struck hard, as the sound will be amplified electronically.

The strings are strung in 18 small fan-shaped clusters, each containing 5 strings. A microphone is mounted at the point where the 5 strings come together. The microphone does not have a membrane, so that the magnet system converts the vibrations of the strings directly into an alternating current of varying power. The electric signal is processed by a system consisting of a series of capacitors, resistors, and coils. The circuit acts as a filter, blocking certain frequencies and allowing others through. This electric filter can be adjusted to achieve the desired sound. After being filtered, the signal is amplified in a valve amplifier and transmitted through the loudspeaker.

The Neo-Bechstein was a very versatile instrument. The volume could be regulated in 12 steps, but if the loudspeaker was switched off, it sounded like a spinet. The left pedal could be used to regulate the volume more precisely. The right pedal had the same function as with a normal piano, sustaining a tone until it faded automatically. A second row of dampers, operated by a separate lever, could make the instrument sound like a reed organ or a harmonium. Because the Neo-Bechstein already had an amplifier and loudspeaker, it was relatively easy to fit connections for a radio receiver and a record player.

Carl Bechstein himself was, of course, full of praise for the versatility of this innovative grand piano: "You can play it for hours, any time of the day and night, without disturbing the neighbors, listen to the latest news reports, or hear Lamond play one of Beethoven's sonatas and immediately try to copy him …"

On August 25, 1931, Bechstein presented the new grand piano with a performance of a Beethoven sonata. The headlines were unanimous in their enthusiasm: "A revolution in piano-building," "A versatile grand piano," or "The electric Beethoven." The press was particularly full of praise for the quality of the highest tones. Nernst, who did not consider himself very musical, responded by saying: "My friend Einstein, who, you know, is very musical, says they sound like porcelain getting smashed."

References

Fritz Wilhelm Winckel, 1931. 'Das Radio-Klavier von Bechstein-Siemens-Nernst, Klangfarben auf Bestellung'. *Die Umschau, Illustrierte Wochenschrift über die Fortschritte in Wissenschaft und Technik* 35 (42), 840–843.

Geertruida Luiberta de Haas-Lorentz, 1935. 'Electrische Muziek'. *Natuur & Techniek* 1935 (2), 22–25.

Curtis Roads, 1996. 'Early Electronic Music Instruments: Time Line 1899-1950'. *Computer Music Journal* 20 (3), 20–23.

Hans-Georg Bartels et al., 2007. *Walther Nernst, Pioneer of Physics and of Chemistry*. World Scientific Publishing, 394 pp.

Albert Einstein

Albert Einstein was not only a world
famous physicist, but also a practical
inventor. Together with Leo Szilard,
he designed three alternative refrigerators. The ideas were ingenious,
but none ever found their way into
the kitchen.

Theory of Relativity

At the age of 16, the young Albert Einstein wondered what a light wave would look like if you could run alongside it at the same speed. According to the classic theory of motion, the wave would have to appear motionless. But that rule does not apply to light. The speed of light is the same, everywhere and at all times, 300,000 m/s, irrespective of the speed or position of the observer. As Einstein would show later, that has far-reaching consequences for our understanding of space, time, and matter.

For centuries, Isaac Newton's classic laws of motion successfully explained the motion of objects not only here on Earth, but also of the planets. That changed at the end of the nineteenth century, when James Clerk Maxwell showed that light consists of electromagnetic waves, immediately raising the question about the medium in which light moved. Maxwell, and later Hendrik Antoon Lorentz, solved this problem with the concept of the "ether." They assumed, for example, that the ether moved along with the rotation of the Earth, thus affecting the speed of light. To everyone's surprise the Michelson–Morley experiments proved that the speed of light is the same everywhere on Earth, whether it moves in the same direction as the Earth's rotation or not. The ether, if it existed at all, thus had no effect on the speed of light. It was Einstein's special theory of relativity that explained the seemingly impossible results of these experiments.

One of the main principles of Einstein's new theory was that the speed of light is constant for all observers. Einstein also stated that the laws of physics are the same for all observers. This led to a number of remarkable conclusions that seem to fly in

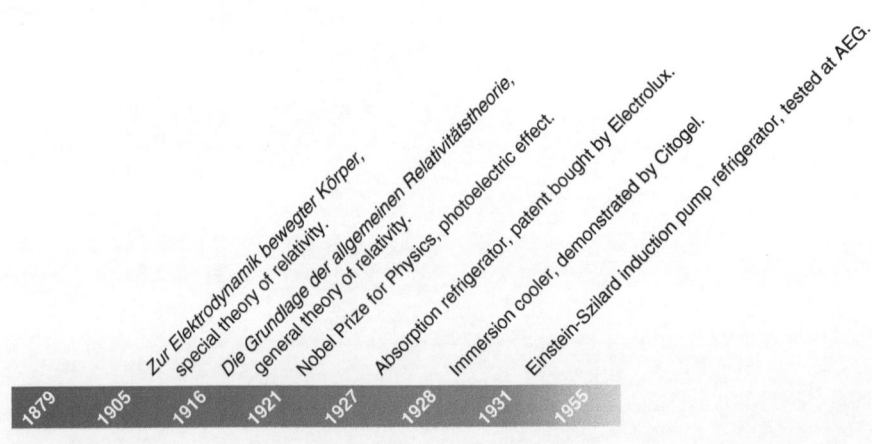

Albert Einstein

the face of all logic. In a moving object, space, time, and mass change, depending on the speed. The object becomes shorter, time goes more slowly, and mass increases. These relative effects increase as the object approaches the speed of light. In our daily lives, we notice little of these effects, as the speeds we experience are only a tiny fraction of the speed of light. Nevertheless, extremely accurate atomic clocks have shown that, in an aircraft, time passes more slowly than on Earth. The most convincing experimental evidence comes from measurements of elementary particles that move only a fraction less than the speed of light.

Because of a disagreement with one of his professors, Heinrich Weber, Einstein was unable to secure a place at a university after completing his studies. He went to work at the Swiss patent office in Berne and developed the special theory of relativity in his spare time. After his groundbreaking ideas were published in 1905, he received one offer after the other from universities throughout Europe.

One of the questions Einstein had as yet been unable to answer at that time was how gravity fitted in with the theory of relativity. Unlike Newton, who assumed that gravity was a force, Einstein thought that it was caused by distortion of the space–time continuum, and was therefore a geometric phenomenon. In 1916, he completed his general theory of relativity, which says that time and space are distorted under the influence of mass. In other words, the presence of mass determines the movement of other objects, and that of light. As with the special theory of relativity, the consequences of the general theory of relativity are difficult to imagine. The effect of mass on the space–time continuum is comparable to that of a marble lying on an elastic sheet: the heavier the marble, the more the sheet is curved.

Refrigerator

"There must be a better way," Albert Einstein thought after reading a tragic story in the newspaper. A family in Berlin had died after being poisoned by toxic gases from a leaking refrigerator. In the 1920s, traditional iceboxes were gradually replaced by mechanical refrigerators. The refrigerants used at the time – methyl chloride, ammonia, and sulfur dioxide – were used in such quantities that a leak could be fatal.

Together with his former student and close friend Leo Szilard, Einstein devised a number of alternative designs for refrigerators. Einstein and Szilard believed that the real problem was not caused by the use of toxic refrigerants, but lay in the moving parts used in conventional refrigerator design. This made leaks almost inevitable. They therefore sought to design systems without moving parts.

Einstein and Szilard met in Berlin in 1920. Einstein was already world famous, thanks to his theory of relativity, while Szilard was just starting out on his scientific career. He could certainly use the income from successful patents. Einstein was also interested in working together and, in the winter of 1925/1926, he signed a business agreement with Szilard, in which Einstein's experience at the patent office proved very valuable. They agreed that all inventions related to refrigeration would be their shared intellectual property. If they made a profit, Szilard would benefit first, as long

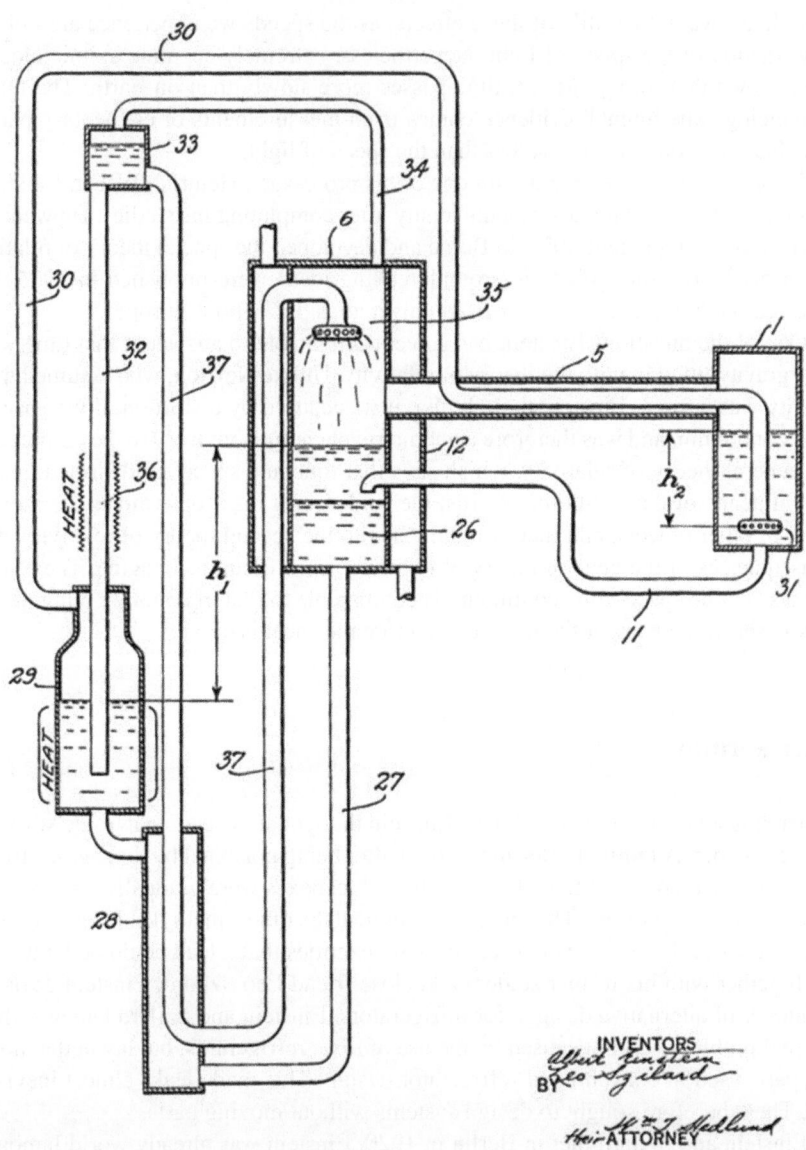

INVENTORS
Albert Einstein
BY Leo Szilard

Hir ATTORNEY

The principle of Einstein and Szilard's absorption refrigerator. Butane evaporates in the cooling section (*1*), under presence of ammonia. The gas mixture of butane and ammonia passes through a pipe (*11*) to the condenser (*6*). In the condenser, water absorbs the ammonia from the gas mixture, releasing the butane. Because liquid butane is lighter than the liquid water and ammonia mixture, it rises to the top of the condenser, where it overflows (*5*) back to the cooling section. The water and ammonia mixture passes through a pipe (*27*) to the generator (*29*) where it is heated and the ammonia evaporates. The ammonia then returns to the cooler through a pipe (*30*), and the water to the condenser through a different pipe (*32*)

as his regular income was lower than that of a university assistant. If not, they would share the profit equally.

During the 7 years in which they worked together, Einstein and Szilard developed three different designs for refrigerators, on which they took out 45 patents. The models were based on the principles of absorption, immersion, and electromagnetism. Although the three principles were substantially different, they were based on a common idea: they had no moving parts and were completely sealed systems.

The absorption system is based on the principle that liquids like water absorb certain refrigerants at low temperatures and release them again at higher temperatures. The refrigerant is driven out of the liquid in a generator by the heat of a gas flame. It then flows through a condenser and an evaporator, where it absorbs heat from the refrigerator. Absorption refrigerators make no noise and are very reliable.

In October 1926, Szilard wrote in a letter to his brother: "The matter of the refrigeration patents, which I applied for together with Professor Einstein, has now come so far that I feel it is a reasonable time to get into contact with industry." A year later, the Swedish company Electrolux bought two patents from Einstein and Szilard for absorption refrigerators.

The application for an American patent on the absorption refrigerator caused a small commotion. An employee of the American patent office wrote back to ask if Einstein was the same Albert Einstein who had devised the theory of relativity. If that were the case, they would have no objection to Einstein's unusual claim that he was both a German and a Swiss citizen.

Albert Einstein together with Leo Szilard in 1946, in a reconstruction of them writing to President Roosevelt in 1939 recommending that America develops an atom bomb before Germany

Einstein and Szilard's patents were actually not for the invention of the gas absorption system, but for an improvement to it. The honor of inventing the system goes to two Swedish engineers, Baltzar von Platen and Carl Munters, who developed a gas absorption refrigerator in 1922. The Platen–Munters cycle used ammonia as the refrigerant and water as the absorption agent. Hydrogen was used in the evaporator to lower the partial pressure of the ammonia, in a similar way to the expansion valve in traditional refrigerators.

Einstein and Szilard's refrigerator used water, ammonia, and butane, and they claimed that it should be able to work with a broader range of temperatures. In their design, butane was the refrigerant and ammonia played the same role as hydrogen in the Platen–Munters cycle.

It is not certain whether Electrolux actually bought the patents to produce the refrigerators, or to protect their own Platen–Munters technology. Either way, the company never built a refrigerator using the Einstein–Szilard cycle. The Platen–Munters system was developed further and is still used on a small scale, for example, in refrigerators for campers.

In the same period, Szilard and Einstein also developed a small immersion cooler, which could simply be dipped into the liquid that had to be cooled. All it required was running water from a tap. The flow of water produced a vacuum in a chamber, in which water, mixed with methanol, would evaporate. The Citogel company demonstrated the cooler at the Leipzig Fair in 1928. The invention worked well, but had little chance of commercial success, as the costs of methanol were higher than expected. The main obstacle, however, was the unreliable water supply, with pressure varying not only between buildings, but also between different floors.

The third design, later known as the Einstein–Szilard pump, was undoubtedly the most revolutionary of the three. A variable electromagnetic field was used to set a metal fluid in motion. The liquid metal acted as a sort of piston, exerting pressure on the refrigerant liquid. In the first instance, the design was intended as an electromagnetic conduction pump in which an electrical current would flow through the liquid metal. Szilard first considered using mercury, but its low conductivity would reduce the efficiency. He then devised a system using a mixture of potassium and sodium, but that was unfeasible because of the aggressive nature of the liquid. The salts would damage the insulation material around the electrical wiring.

Szilard consulted Einstein about the problem, who suggested using electromagnetic induction, making wires unnecessary. The force was instead transferred to the liquid metal by external coils. Although this system was less efficient than standard compressors, it was allegedly safer. In 1928, the German company Allgemeine Elektrizitäts-Gesellschaft (AEG) agreed to build a prototype. Szilard and a number of others were employed by the company to develop the refrigerator further.

In 1931, an Einstein–Szilard refrigerator was subjected to a 4-month test at AEG's research institute. It proved to work efficiently, but made a lot of noise. Descriptions by witnesses varied from the sound of running water to the howl of a jackal.

Dark clouds were, however, gathering over the project. The development of conventional refrigerators was advancing much more rapidly. The invention of freon, which was believed not to be toxic, considerably reduced the danger from leaks and, as yet, no one was aware of its harmful effects on the atmosphere. The spread of Nazism was also a cause of concern to the two researchers. After the Nazi party won the elections in 1930, Szilard wrote to Einstein expressing his concerns about developments in Europe, and his doubts about ever being able to build a refrigerator with an electromagnetic pump there. "From week to week I detect new symptoms, if my nose doesn't deceive me, that peaceful developments in Europe in the next 10 years is not to be counted on … . Indeed, I don't know if it will be possible to build our refrigerator in Europe."

Lastly, AEG did not remain impervious to the economic depression and, in 1932, the project was stopped. Some months later, Adolf Hitler became German chancellor and Szilard and Einstein fled abroad. Like their other two refrigerators, the models with an Einstein–Szilard pump never made it to the kitchen, but the principle is still applied today – in the cooling of nuclear reactors.

References

Gene Dannen, 1997. 'The Einstein-Szilard Refrigerators'. *Scientific American* 1997 (January) 74–79.

Sam Shelton et al., 1999. 'Second Law Study of the Einstein Refrigeration Cycle'. *Proceedings of the Renewable and Advanced Energy Systems for the 21st Century*, 1–9.

Harlow Shapley

Harlow Shapley refuted the idea that the Sun is at the center of the Milky Way. On Mount Wilson, where his telescope was located, he observed not only the stars, but the ants. He conducted remarkable research into how the speed at which they moved was affected by the ambient temperature.

R. Schils, *How James Watt Invented the Copier: Forgotten Inventions of Our Great Scientists*, DOI 10.1007/978-1-4614-0860-4_21, © Springer Science+Business Media, LLC 2012

The Center of the Milky Way

Harlow Shapley originally wanted to be a journalist. He had already gained some experience with a local newspaper and was on the point of continuing his studies in journalism. He went to enroll at the University of Missouri, but found he could no longer be admitted. Not wishing to return home without achieving something, he decided to choose a course of study that he could start immediately. According to his own account, he started at the top of the alphabetical list in the course directory. The first subject, archeology, did not appeal to him, but the second, astronomy, sounded more promising.

After graduating in astronomy in 1911, he went to Princeton to do his thesis under the supervision of Henry Norris Russell, famous for the Hertzsprung–Russell diagram. The diagram shows the relationship between the luminosity and temperature of a star. It is an excellent instrument for classifying stars and helps understand their evolution. Shapley analyzed Russell's data on more than 90 eclipsing binary stars. These are twin stars that rotate around each other and eclipse each other in turn, so that, seen from the Earth, they display a fluctuating luminosity. Russell and Shapley developed a method of estimating the size of binary stars on the basis of their luminosity.

After he had completed his thesis, Shapley's attention shifted to globular clusters, collections of tens to hundreds of thousands of stars. In Shapley's time, these

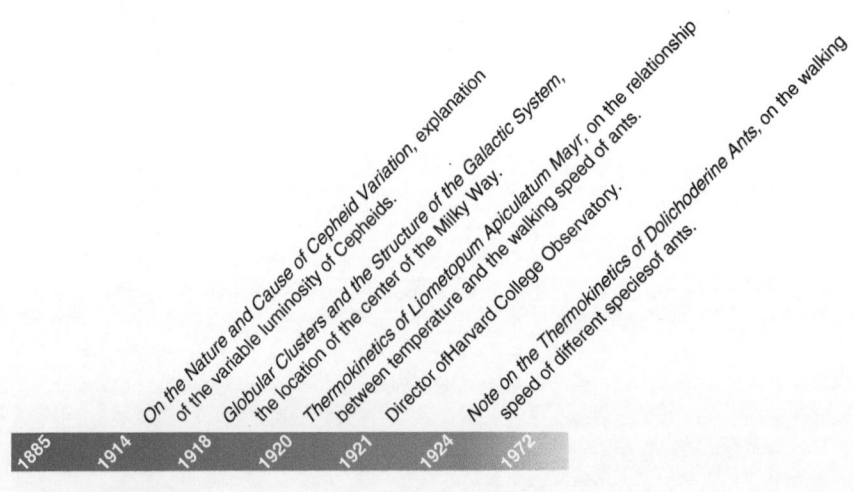

Harlow Shapley

clusters were known to be distributed asymmetrically, and to contain many Cepheids, stars whose luminosity varies over periods ranging from less than a day to around 100 days. Because of the direct correlation between their absolute luminosity and pulsation periods, Cepheids are very useful in determining distances in the universe.

To research globular clusters, Shapley worked at the Mount Wilson Observatory from 1914 to 1921. The observatory, standing 1,700 m above Los Angeles, was built in 1904 and had one of the largest telescopes in the world at the time. Using the data from several hundred newly discovered Cepheids, Shapley refined the method of determining cosmic distances, and used it to chart the locations of globular clusters. He observed that a third of the hundred known clusters were to be found in the direction of the constellation of Sagittarius.

Because the clusters were globular in shape, Shapley correctly assumed that they were located at the center of the Milky Way. He calculated that the Sun was 50,000 light years from the center. Although he considerably overestimated this distance – now known to be 30,000 light years – his discovery completely turned the existing image of the solar system on its head.

Until 1918, the view of the solar system was largely based on the work of Dutch astronomer Jacobus Kapteyn, who thought that the Sun lay roughly at the center of a cluster of stars in the shape of a lens. It was difficult to establish the borders of the system, but it was estimated to be some tens of thousands of light years in size.

Shapley's dethronement of the Sun as the center of the Milky Way has been compared with Copernicus' displacement of the Earth as the center of the solar system. In a purely astronomical sense, however, Shapley's discovery was more remarkable for initiating a new phase in our understanding of the spectacular dimensions of the solar system and the universe.

Ants

Harlow Shapley spent 7 years at Mount Wilson Observatory. It was a daunting task to determine the periods of the Cepheids. In a letter to Jacobus Kapteyn on February 6, 1917, he wrote: "The work on clusters goes on monotonously …" He hated the cold of the mountain: "I suffered quite a bit those long, cold nights. I suppose I didn't get as much sleep in the daytime as I needed, for I was running around observing ants in the bushes."

Shapley liked to walk in the area surrounding the observatory and observe everything he came across. One day, he saw a trail of ants walking back and forth across a concrete wall. At one point, a side trail branched off from the main track and disappeared into the shadow of some bushes. Shapley noted that the ants in the shadow visibly slowed down. This surprised him so much that he wanted to learn more. The next time, he took a thermometer, barometer, hygrometer, ruler, and stopwatch along. He recorded the speed of the ants over a distance of 30 cm at various times of the day and night, also carefully noting the temperature, air pressure, and humidity.

The ants proved to be the fastest in the midday sun, slowing down at the end of the day and walking the slowest at night. Shapley was the first to observe the thermokinetic properties of ants. Their speed increased as the temperature rose, but was unaffected by changes in air pressure, humidity, or season. He described his findings in a paper and showed it to astronomer and editor Frederick Seares. Seares glanced at the title, laughed, and pushed it back across the table to Shapley. It must have been another of Harlow's jokes.

But it was no joke. Shapley's discovery was scientifically important because it clarified variations in activity among cold-blooded animals. That activity – in this case, their capacity to walk – depends on the metabolic processes in the ant's body. Temperature determines to a significant degree the speed of these processes and, consequently, the speed at which the ants walk. According to Shapley, that may have been described qualitatively but, until now, it had not been backed up by quantitative data. With his observations on Mount Wilson, Shapley saw a unique opportunity to fill this gap.

Together with growth and subsistence, motion accounts for the largest part of an ant's energy consumption. In an ecological sense, the speed with which they move is important to the size of the area in which they can find food, evade predators, and adapt to changes in their environment.

Shapley conducted his observations on the species *Liometopium apiculatum*, which is widely prevalent in the southwest of the USA. Shapley considered this

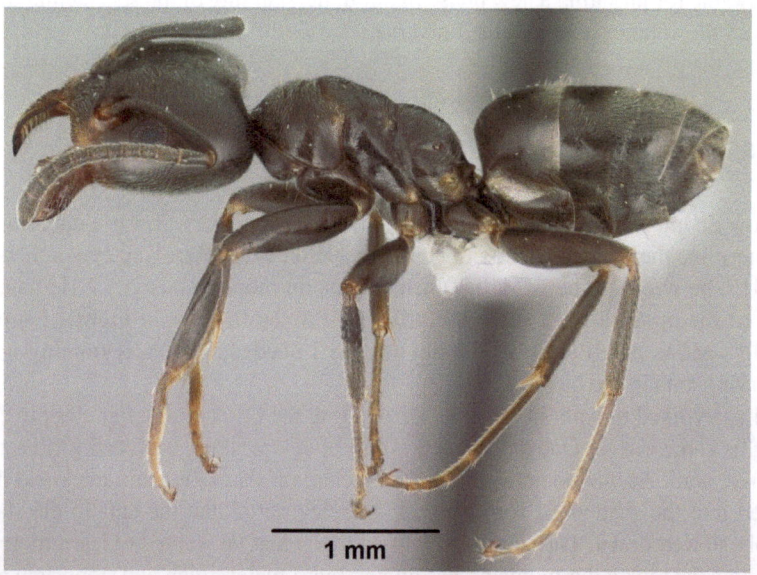

The ant species that Harlow Shapley studied is now known as *Liometopum luctuosum*. In Shapley's time, they were still thought to be a subspecies of *Liometopum apiculatum*

species exceptionally suitable for this research because they follow fixed routes, which they use for long periods. During his observations on Mount Wilson, the ants used the same trail for the whole period of 2 years. That made it easy for Shapley to set up two permanent observation sites, one half a meter from the nest and the other at a distance of 15 m. An additional advantage of a fixed trail is that the ants have a stable energy consumption, energy that they mainly use to move.

The ants were active both day and night, in temperatures ranging from 8 to 38°C, enabling Shapley to study their walking speed in a wide range of conditions. The large populations in the nests made it easy for him to gather substantial data for statistical analysis. In one nest, as many as 70,000 ants could pass in and out in a day. In the warmer months, around a 100 ants a minute would pass through his speed traps.

Shapley's measurements soon showed that temperature was the main factor determining the ants' walking speed. He deduced a reliable empirical curve, which showed that, at a temperature rise of 30°C, the speed increased from 0.44 to 6.6 cm/s.

Shapley claimed that the correlation was so strong that you could calculate the temperature to an accuracy of 1°C from the average walking speed of 10–20 different ants. He even found that the ants' walking speed responded more quickly than a mercury thermometer to a sudden change in temperature.

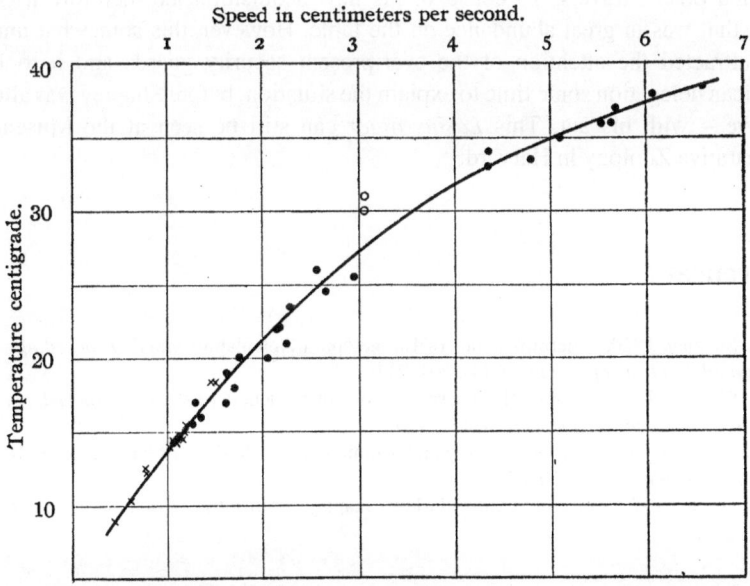

The relationship between the walking speed of *Liometopum apiculatum* and temperature, as observed by Shapley on Mount Wilson in 1920. The *closed dots* are the observations made at the test site half a meter from the nest. The *crossmarks* are from the site at 15 m distance. The two *open dots* show observations where the temperature was measured incorrectly

Outsiders may have considered Shapley's digression into biology as little more than a hobby, but he always felt that it was equally important as the rest of his scientific career. In a letter to a science journalist in 1920, he wrote: "There is nothing of particular interest in my career, ... In 1914 I joined the staff at this observatory and have devoted myself mainly to stellar photometry standard, eclipsing and Cepheid variables, star clusters, ... entomological physiology, and of late to the application of the intensifier to nebular problems." He did not conceal his admiration for ants. Once, during a lecture, he exclaimed: "When you go out of your way to step on an ant, you insult the order of nature, for you, a mere social upstart, are jumping on a creature that perfected a social system some 30,000,000 years ago!"

After Shapley ended his successful period at Mount Wilson, he became director of Harvard College Observatory. Nevertheless, he found time to publish a sequel to his first study. One of his findings was that the correlation he had observed between temperature and walking speed also applied to other ant species than *Liometopium apiculatum*.

Shapley's love of ants did not wane as he grew older. In 1945, it even led to a minor international incident. To mark the 220th anniversary of the Scientific Academy of the Soviet Union, the guests were treated to an elaborate banquet attended by Joseph Stalin. During the long dinner, Shapley noticed an ant having a feast of its own in the fruit bowl on the table. He caught the ant and put it in a small glass bottle that he always carried with him for such eventualities. Shapley had a habit of using the strongest drink available in the country he was visiting to anesthetize the ants he collected on his travels. To conserve his new acquisition, he therefore used the vodka that was in great abundance on the table. However, this somewhat unusual deed attracted the attention of the ever-present security guards and it took the American delegation some time to explain the situation, before Shapley was allowed to leave – with his ant. This *Lasius niger* can still be seen at the Museum of Comparative Zoology in Harvard.

References

Harlow Shapley, 1920. 'Thermokinetics of Liometopum Apiculatum Mayr'. *Proceedings of the National Academy of Sciences* 6 (4), 204–211.
Harlow Shapley, 1924. 'Note on the Thermokinetics of Dolichoderine Ants'. *Proceedings of the National Academy of Sciences* 10, 436–439.
Owen Gingerich, 1973. 'Harlow Shapely and Mount Wilson'. *Bulletin of the American Academy of Arts and Sciences* 26, 10–24.
Zdeněk Kopal, 1972. 'In memoriam: Harlow Shapley'. *Astrophysics and Space Science* 18, 259–266.

Erwin Schrödinger

Schrödinger's wave equation, one of
the founding principles of quantum
mechanics, earned him the Nobel
Prize in 1933. Ten years later, a short
venture into biology resulted in
Schrödinger's personal bestseller
What is life?, in which he foresaw the
physical–chemical basis of life, long
before the discovery of DNA.

R. Schils, *How James Watt Invented the Copier: Forgotten Inventions
of Our Great Scientists*, DOI 10.1007/978-1-4614-0860-4_22,
© Springer Science+Business Media, LLC 2012

Wave Equation

In the fall of 1906, Erwin Schrödinger began studying physics at the University of Vienna, where he was taught by Franz Exner and Fritz Hasenöhrl. From Exner, he mainly learned about experimental physics, while Hasenöhrl set him on the trail of theoretical physics, a road he would follow for the rest of his life. Schrödinger's intended university career was disrupted by the outbreak of the First World War. On July 28, 1914, Austria entered the war and, a few days later, Schrödinger was mobilized.

He spent 3 years on the Italian front with the artillery. As a theorist, he could fortunately work with just pen and paper, and he succeeded in producing scientific publications regularly throughout the war. In this period, Schrödinger became acquainted with Albert Einstein's general theory of relativity. He was not the only one to be impressed; back at the university in Vienna, his colleagues were all just as excited about Einstein's work. Toward the end of the war, Schrödinger wrote his first paper on quantum theory. Although it only summarized existing theories and contained no original ideas, it clearly marked Schrödinger's transition to a new area of scientific activity.

Some years later, Schrödinger took the opportunity to work at the University of Zurich, where illustrious researchers like Albert Einstein and Max von Laue had preceded him. The Zürich years would become Schrödinger's most successful. In 1926, he published a series of six papers on his groundbreaking work on wave mechanics.

From as long ago as the seventeenth century, there had been various theories on the nature of light. According to Christiaan Huygens, it consisted of waves, but

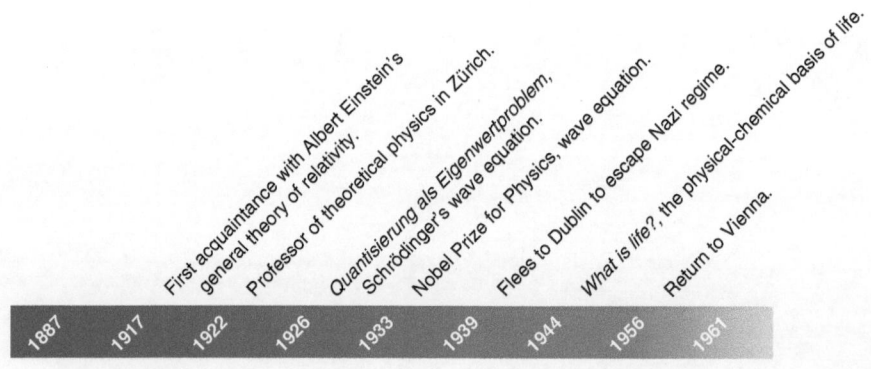

Erwin Schrödinger

Isaac Newton was the first to describe it as being made up of particles. A first step toward reconciling this apparent contradiction – the wave–particle duality – was taken by Louis de Broglie who suggested in his thesis that all matter, irrespective of its size, had an associated wave.

Inspired by De Broglie's thesis, Schrödinger set about finding the wave functions that described changes in the behavior of matter in time and space. The wave equation that bears his name is a differential equation that can be used to calculate how the wave function of a particle changes under the influence of external forces. Schrödinger successfully applied his findings to explain the properties of the hydrogen atom. In quantum mechanics, Schrödinger's wave equation fulfills the same function as Newton's laws of motion in classical mechanics.

What Is Life?

Around the end of the 1930s, the Irish Prime Minister Eamon de Valera was toying with the idea of establishing a leading institute of physics in Dublin. His idea was based on the Institute for Advanced Studies in Princeton, where Albert Einstein found refuge after fleeing Nazi Germany. The Dublin Institute for Advanced Studies

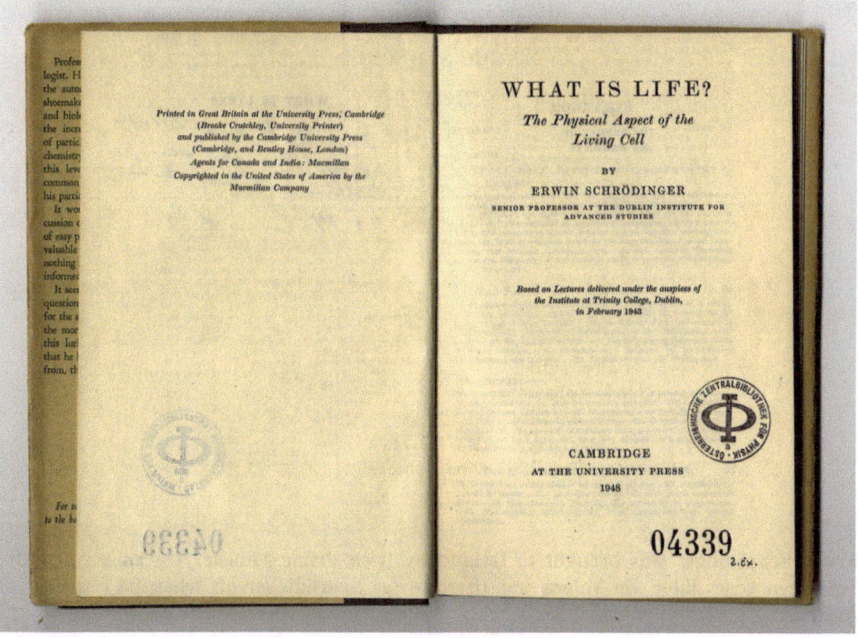

Title page of the third edition (1948) of Erwin Schrödinger's *What is Life?*

Farewell meeting

Mr. de Valera with Professor and Mrs. Schroedinger at Leinster House yesterday.

NAME OF IRELAND NOT IMPORTANT

—Commons Speaker

LONDON, Thursday—Mr. Harry Hynd, Labour M.P. for Accrington, asked the Speaker in the British House of Commons to-day whether a question from Captain L. P. S. Orr, Unionist M.P. for South Down, should not have referred to the Government of the Republic of Ireland, and not the "Eireann Government."

The Speaker replied that "so long as the sense is intelligible, the words are not a very important matter."

Capt. Orr asked the President of the Board of Trade, Mr. Peter Thorneycroft, if the import levies imposed by the "Eireann" Government were in accordance with the Anglo-Irish trade agreement.

Mr. Thorneycroft — These recent moves were dictated by "Eire's" balance of payments position, and, so far as I can judge, she has made a real attempt to keep within the terms of the agreement.—I.N.A.

'Support railways' call in Dundalk

Members of the G.N.R. Employees Association, formed recently to get more business for the railways and so prevent the closing branch lines, have been distributing leaflets and canvassing traders in Dundalk.

The leaflets point out that the G.N.R. employs in Dundalk 1,850 men, who last year received £761,000 in wages, most of which was spent in the town. It is also pointed out that the G.N.R. paid £5,758 in rates to the Dundalk Urban Council and £242 to the Louth Co. Council.

A member of the Association said yesterday that they were satisfied with the results of their canvass so far.

Serov talks with Scotland Yard

Colonel General Ivan Serov, head of the Soviet State security, who arrived in London by air yesterday, had his first conference with Scotland Yard officers on the security plans for the visit to Britain next month of the Soviet leaders, Marshal Bulganin and Mr. Khrushchev.

He is expected to return to

Dr. Shroedinger leaves Ireland

MR. de VALERA BIDS SCIENTIST FAREWELL

IRISH PRESS Reporter

PROFESSOR ERWIN SCHROEDINGER, world-famous scientist, left Ireland for his native Vienna last night. He sailed from Dublin on the B. and I. m.v. Leinster, accompanied by his wife, Mrs. Annemarie Schroedinger. And the last person to shake his hand in farewell was Mr. de Valera, who, 17 years ago, invited him to Ireland.

One of our age's great men of science

WHEN Erwin Schroedinger sailed from Dublin last night, writes a special correspondent, we said good-bye to one of the great scientists of our age. A notable figure of the revolution in scientific thought, which was led by Planck and Einstein, he will be remembered with these immortals. Not that his contributions to knowledge were merely developments of theirs — they were as original and in many ways as far-reaching.

It is not easy to convey a complete picture of the man. A vivid personality with an acute mind and courteous manner, he likes controversy. The inquiring scientist and mathematician in him does not stifle an interest in humanity and literature, nor those poetic qualities that go with great imagination. The power of intellect coupled with a capacity for dreams—these are the qualities of genius, and Schroedinger possesses them.

Atomic theory

When, in the 1920's, he gave the world his wave mechanics, he made a fundamental contribution

Mr. de Valera went aboard just before the boat sailed. Professor Schroedinger, who was indisposed, had retired on going aboard. Mr. de Valera went to his cabin, and few words were spoken as the old friends exchanged an eloquent handshake.

Major Vivion de Valera, T.D., also went aboard to bid farewell to the Professor.

Mr. de Valera bade good-bye to Mrs. Schroedinger in the lounge as he left the boat.

Yesterday Mr. de Valera and Professor Schroedinger met in Leinster House and talked for 30 minutes. They first met in Geneva in 1939, and that meeting was recalled by the professor yesterday.

"Mr. de Valera was most kind," he said, "and invited us to Ireland and helped us to get here. We enjoyed our stay in Ireland and made many friends and were most happy."

His wife said: "We were so very happy here. We are too excited yet to feel sorry we are leaving, but I know we will feel sorry."

"But we are going back to our own home country, and you know when one gets older one loves to be going back. Vienna is our home."

They will spend some days in London and Innsbruck before going on to Vienna, where the Professor takes up an appointment at Vienna University.

Among those who saw them off was Professor Lanczos, of the Institute of Advanced Studies.

Erwin Schrödinger was brought to Ireland by Irish Prime Minister De Valera in 1939. Seventeen years later, De Valera was there to see Schrödinger off when he returned to Austria

was soon established but, to complete the picture, it needed to attract one of the world's top names in physics.

A unique opportunity soon presented itself when De Valera heard that Erwin Schrödinger was under heavy pressure in his homeland. His declared opposition to Nazism had made his work at the University of Graz, where the rector was a Nazi, almost impossible. After the *Anschluss*, Schrödinger was forced to leave Austria and seek refuge elsewhere. At that moment, because of his pioneering work in quantum mechanics, he could justifiably be called a leading figure in physics.

Consequently, in 1939, the Nobel Prizewinner arrived in Ireland, where he spent 16 years trying to explain all the fundamental forces of nature in a single unified field theory. Although he failed in this ambition, he did succeed in making Dublin an internationally renowned center of theoretical physics. In that period, Schrödinger himself produced around 50 scientific publications.

In 1943, Schrödinger gave the annual Trinity College public lecture. He deliberately avoided the more obvious topics of wave mechanics and electromagnetic fields and chose something completely different: "a naïve physicists' approach to the phenomenon of life." Schrödinger had earlier been alerted to the paper *Über die Natur der Genmutation und der Genstruktur* (On the nature of gene mutation and gene structure) by Timoféeff-Ressovsky, Zimmer, and Delbrück, which suggested for the first time that gene mutation is caused by a change in a single location in a molecule. Schrödinger saw in the discontinuous way in which mutations occur, a strong similarity with quantum mechanics. It inspired him to devote a series of three public lectures to his ideas on how heredity is determined by chemical and physical mechanisms.

These popular scientific lectures formed the basis of the book *What is life?*, published in 1944 with the subtitle *The Physical Aspect of the Living Cell*. Totally against all expectations, the book sold like hot cakes, with more than 100,000 copies going over the counter in a short time.

Schrödinger starts the book by asking how the events that occur in a living organism in space and time can be explained by physics and chemistry. He says that, although these disciplines did not, at that time, offer satisfactory explanations, that by no means suggests that they cannot be used to explain life processes. In his view, the road to understanding life starts with the awareness that it is based on purely mechanical actions. That implies that a biological system can be completely described and analyzed by mathematical equations.

Schrödinger writes that, in some way or another, chromosomes contain the complete code for the development of an individual and that the phenotype, the "manifest nature of the individual," can be completely predicted from the code-script. Particularly striking is the claim that the gene is a kind of aperiodic crystal, or a chain of various recurring units. He compares the different units with the dots and dashes of the Morse code.

Despite its high sales, the book did not escape criticism from scientific quarters. Schrödinger thought that the gene was a protein in which each atom, radical and heterocyclic, ring played an individual role. Because he was not a biologist himself, he had to rely on the work of others, and drew particularly heavily on earlier work by the three-man ship Timoféeff-Ressovsky, Zimmer, and Delbrück. In fact, his description of the gene is simply a reformulation of Delbrück's suggestion that a gene is a polymer, built up of recurring identical structures. But this was not the only criticism of the book. The content was already obsolete when it was published, because Schrödinger had unfortunately spoken to the wrong biologists, who still believed that genes consisted of proteins. Some months previously, Oswald Avery had discovered that genes consist of DNA, but Schrödinger was not yet aware of that. He had also been unaware of other contemporary developments, such as the use of phages in pursuing the structure of DNA. Max Perutz, winner of the Nobel Prize for Chemistry, summed up the criticism by saying: "What was true in the book was not original, and most of what was original, was not true."

If the book contained so little that was new, why did it sell so well and is still considered a groundbreaking work? First of all, it is clear and accessible, for both the scientist and the interested lay reader. Second, the timing was perfect. Driven by Max Planck, Albert Einstein, Niels Bohr, Werner Heisenberg, and, not in the last instance, Schrödinger himself, modern physics had really taken off and there was great confidence that it would exert a considerable influence on all scientific disciplines, including biology.

On both sides of the Atlantic, reputed scientists were working to develop the atomic bomb or on other war-driven technologies. While the rest of the scientific world was trapped in military research, in Dublin Schrödinger could practice science freely. With *What is life?*, he gave researchers around the whole world the meaningful prospect of placing modern physics once again in a favorable spotlight.

For many biologists, Schrödinger's name is forever associated with *What is life?* After reading the book, James Watson decided to devote all his energies to unraveling the structure of DNA. Fellow Nobel Prize winners Francis Crick and Maurice Wilkins were also unanimous in their praise for *What is life?* The more recent commendations emphasize that the book may not have offered any readymade answers to the question in the title, but it did suggest a new direction for research, a new way of addressing the essential questions in biology. This places the book at the cradle of molecular biology.

References

Erwin Schrödinger, 1944. *What Is Life? The Physical Aspect of the Living Cell*. Cambridge University Press, 92 pp.

Neville Symonds, 1986. 'What Is Life?: Schrödinger's Influence on Biology'. *The Quarterly Review of Biology* 61 (2), 221–226.

Walter Moore, 1994. *A Life of Erwin Schrödinger*. Cambridge University Press, 363 pp.

Enrico Fermi

By generating the first controlled nuclear reaction, Enrico Fermi was present at the dawn of the nuclear age. During a lunch with colleagues, he asked lightheartedly "Where is everybody?", setting off a prolonged discussion on extraterrestrial life.

R. Schils, *How James Watt Invented the Copier: Forgotten Inventions of Our Great Scientists*, DOI 10.1007/978-1-4614-0860-4_23,
© Springer Science+Business Media, LLC 2012

Nuclear Reaction

After completing his education in physics, Enrico Fermi worked with some of the leading figures in quantum mechanics, including Max Born in Germany and Paul Ehrenfest in the Netherlands. In 1926, with Paul Dirac, he developed a theory on the statistical behavior of electrons that comply with the "Pauli principle," a basic quantum mechanical principle that states that no two electrons can occupy the same state simultaneously. Today, all particles that comply with the "Fermi-Dirac statistics" – including electrons, protons, and neutrons – are known as fermions.

A year later, Fermi was appointed Professor of Theoretical Physics at the University of Rome. It was an exciting time in physics, with one breakthrough after another. In 1932, the neutron was discovered in England and, 2 years later, French researchers succeeded in generating artificial radioactivity by bombarding elements with helium nuclei. Fermi came up with the idea of using neutrons instead of helium nuclei and discovered that the use of slow neutrons was a very effective method of generating nuclear transformations. With neutron bombardment, he succeeded in changing the number of neutrons or protons in atomic nuclei, creating other elements and isotopes.

The kind of radioactivity released during these tests provided Fermi with information on the reactions that had taken place. Bombarding uranium released a different kind of radioactivity than he had expected. Fermi had no immediate explanation for this, but suspected that he had succeeded in adding a proton to the uranium nucleus. With 92 protons, uranium is the heaviest element that occurs naturally on

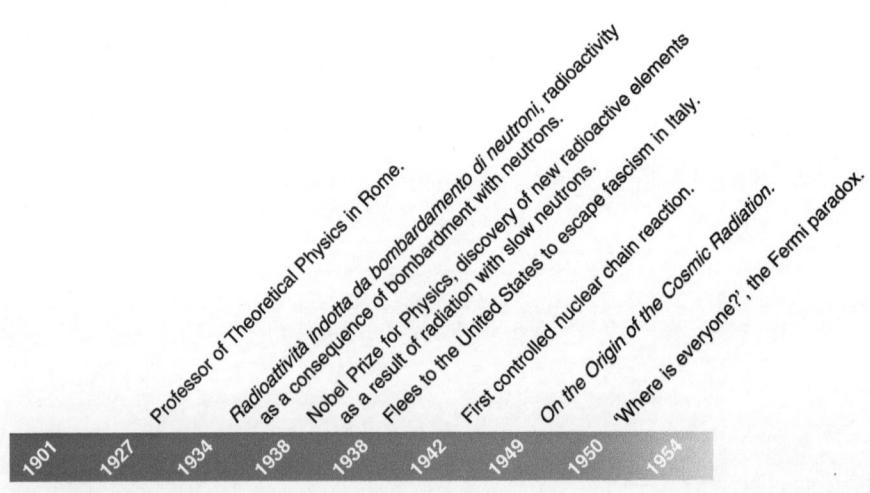

Enrico Fermi

Earth and the quest for "transuranics" was a goal in itself. The title of his paper, *Possible Production of Elements of Atomic Number Higher than 92*, shows that he was not so sure of himself.

Fermi was awarded the Nobel Prize for Physics in 1938 for his experiments with neutrons. After the presentation in Sweden, Fermi decided to flee the fascist regime in Italy and migrated to the USA, where he was welcomed with open arms. Some years later, Otto Hahn, Lise Meitner, and Fritz Strassmann repeated Fermi's experiments in Germany and made the astounding discovery that Fermi had been wrong in assuming that he had created a transuranic element. On the contrary, the uranium atom had split into lighter elements. The uranium used contained 92 protons and 143 neutrons. The bombardment with neutrons created a new isotope of uranium with one extra neutron. But this new isotope proved highly unstable and separated into barium, with 56 protons and 85 neutrons, and krypton, with 36 protons and 56 neutrons. It was Meitner who realized that, when uranium separated, a great deal of energy was released by the decrease in the total mass.

In the USA, Fermi now recognized that, without being aware of it, he had generated the first artificial nuclear fission reaction in 1934. He was fully aware of the great importance of Meitner's analysis and, together with Niels Bohr, conceived the idea that it must be possible to set a nuclear chain reaction in motion. After all, when uranium decays, another three neutrons are released, which can in turn fuse with other uranium nuclei.

Like others, Fermi also saw the danger posed by this new development and he warned US President Roosevelt that the Germans might use it for military purposes. To beat Germany to it, Roosevelt immediately initiated the Manhattan Project, with the aim of developing an atomic bomb.

Within the project, Fermi was responsible for generating the first controlled nuclear reaction. On December 2, 1942, he achieved that goal in a laboratory at the University of Chicago. The technology was then used in the first atomic bombs. The peaceful use of nuclear fission as an alternative source of energy was not developed until after the war.

The Fermi Paradox

In the summer of 1950, Enrico Fermi visited the Los Alamos National Laboratory, where preparations were underway for the "greenhouse tests," the fourth series of nuclear tests after the Second World War. One day, he was walking to Fuller Lodge with Edward Teller, Emil Konopinski, and Herbert York to have lunch and the conversation turned to flying saucers. Some time earlier, a cartoon in the *New Yorker* had depicted aliens abducting trash bins from the streets of New York. The cartoonist had combined two concerns occupying the minds of the people of the city at that time: the unexplained disappearance of trash cans and reports of UFOs. Fermi joked that the cartoon presented a reasonable theory, since it adequately explained two separate phenomena.

The conversation turned to whether flying saucers would in some way or another be able to fly faster than the speed of light. Fermi allegedly asked Teller: "Edward, what do you think? How probable is it that within the next 10 years we will have clear evidence of a material object moving faster than light?" Teller thought that the odds were only one in a million, but Fermi obviously felt differently and estimated the probability at one in ten.

The conversation moved on and, during lunch, everyone seemed to be occupied with other things. Their earlier conversation was, however, still buzzing around in Fermi's mind. Suddenly, he said: "But where is everybody?" We no longer know his exact words as, more than 30 years later, his three table companions all remember it differently. Nevertheless, whatever Fermi said, they all laughed. They knew, with no need of further explanation, that he was referring to extraterrestrial beings.

According to Teller, the conversation did not go much further than a few remarks about the vast distances to the nearest possible inhabited planet, and that the solar system was in a rather unfavorable location, rather like a suburb in a large city. York, however, remembers Fermi being more explicit, quoting figures on the probability of there being planets similar to the Earth, of there being life on them, of these life forms developing advanced technologies, and so on. Whatever was actually said, Fermi's conclusion was that we should long ago have been visited by extraterrestrial

The cartoon in the *New Yorker Magazine* on May 20, 1950, on the link between UFOs and the mysterious disappearance of trash cans

beings, and that it was therefore strange that we have not yet seen any evidence at all of extraterrestrial life.

Fermi himself never committed any of this to paper, but his reasoning was probably more or less as follows: our galaxy is 10^{10} years old and is approximately 100,000 light years from end to end. The time required to colonize a galaxy would depend on the speed with which the colonists could explore space. At speeds of around a thousandth of the speed of light – i.e., around 300 km/s – it would take 10^8 years to explore it. The key to this paradox is that our galaxy is older than the estimated exploration time by a factor of 100. If beings had ever advanced sufficiently to travel through space at the required speed, they would have been able to explore the whole galaxy in a relatively short time. Yet we have never seen them.

It was David Viewing who, 25 years later, named the paradox after Fermi. Fermi was, however, neither the first nor the only scientist to toy with the idea. It should actually be called the Tsiolkovski–Fermi–Viewing–Hart–Tipler paradox. Back in 1933, Konstantin Tsiolkovski had written that mankind should seek its future beyond the Earth: "The Earth is the cradle of humanity, but we cannot live forever in a cradle." Tsiolkovski was a believer in monism, the belief that there is only one

The Milky Way comprises some 200 billion stars, including the Sun

reality that applies to the whole universe. If mankind starts to explore space, other beings will do the same and, sooner or later, we will meet each other.

In the 1970s, Michel Hart and Frank Tipler took the discussion further. Hart was the first to come up with a detailed analysis of the possible solutions, while Tipler adopted the rather extreme standpoint that, if extraterrestrial beings did exist, they would build self-reproducing spacecraft that could explore the universe in a relatively short time.

The first search for extraterrestrial life started in 1960. Frank Drake used a large parabolic antenna to try and pick up radio signals. The search focused on both actively transmitted signals and what is known as "leakage." The Earth, for example, has been leaking signals into space ever since radio – and later, television – broadcasts started.

Drake tried to estimate the number of extraterrestrial civilizations we should be able to contact. The "Drake equation" is based on seven factors, which partly coincide with Fermi's ideas. They include star formation, the fraction of stable stars with planets, the planets that could support life, the fraction of those where intelligent life could develop, and the average lifespan of a civilization. The result is of course extremely dependent on the value given to these factors, so that estimates range from 50 civilizations to as many as 250 million.

Anyone thinking that the discussion on Fermi's paradox is slowly fading away would be wrong. In recent decades, a number of developments have occurred that have breathed new life in to the debate. New planets are regularly discovered orbiting one star or another, and several hundred of these exoplanets have been charted. Scientists are also improving our understanding of what is known as the "Galactic Habitable Zone," the part of the galaxy where life might be able to develop. On Earth itself, microorganisms have been discovered that can live under very extreme conditions, changing our conception of the possibilities of life evolving beyond the Earth.

In the course of time, many solutions to the Fermi paradox have been proposed. Some of these assume that alien civilizations exist but do not explore space or colonize other parts of the universe. This may be because they do not have the technology, or because they are simply not interested. It is also possible that advanced civilizations die out before they have developed the technology required to explore the galaxy.

Other solutions assume that alien civilizations do travel beyond their home planets and have perhaps long visited the Earth, without us knowing about it. Others suggest that advanced civilizations have no interest in us because we are so primitive. Perhaps they have ethical reasons for not contacting us, and just leave us alone. Last, there is the possibility that the evolution of intelligent life is more difficult than we think, and the we are indeed more advanced than all the others.

In addition to those who seek a solution to the paradox in one of these two categories, there are people who claim that, in a strict sense, there is no paradox, because it is not based on sound logic. You cannot conclude that, because you have not observed alien life forms, that they do not exist. We should just make more of an effort to find them. After all, we have only been looking for a short time and have only explored a minimal part of the universe.

As for Fermi, as far as we know, after his comment during lunch, he never said anything more on the matter. However, in the final years of his career, he worked on a theory of the origins of cosmic radiation. It is very unlikely that, while he was working, his thoughts never returned to that lunchtime conversation in Los Alamos.

References

Glen David Brin, 1983. 'The 'Great Silence': the Controversy Concerning Extraterrestrial Intelligent Life'. *Quarterly Journal of the Royal Astronomical Society* 24, 283–309.

Robert Freitas, 1984. 'There Is No Fermi Paradox'. *Icarus* 62, 518–520.

Eric Jones, 1985. *'Where Is Everybody?': An Account of Fermi's Question*. Los Alamos National Laboratory, Los Alamos, 12 pp.

Stephen Webb, 2002. *If the Universe is Teeming with Aliens ...Where is Everybody? Fifty Solutions to the Fermi Paradox and the Problem of Extraterrestrial Life*. Copernicus, New York, 300 pp.

Rosalind Franklin

Rosalind Franklin played a key role in the discovery of the structure of DNA. Her photograph of a DNA molecule led James Watson and Francis Crick to the solution. At the start of her career, Franklin performed groundbreaking research into the structure of carbon.

R. Schils, *How James Watt Invented the Copier: Forgotten Inventions of Our Great Scientists*, DOI 10.1007/978-1-4614-0860-4_24,
© Springer Science+Business Media, LLC 2012

DNA

In 1951, after spending 4 years in Paris, Rosalind Franklin returned to her birthplace, London. In Paris, she had become a specialist in X-ray diffraction, a method excellently suited to determining the structure of solids. Back in London, at King's College, she hoped to use X-ray diffraction to disentangle the structure of DNA.

In the same laboratory, Maurice Wilkins was also studying DNA, but their cooperation was very strained from the outset. John Randall, who had been responsible for Franklin's appointment, had failed to make the division of labor clear: Wilkins thought that Franklin would be assisting him, but Franklin went her own way.

Besides King's College, there were two other institutes in the race to be the first to discover the structure of DNA. At the Cavendish Laboratory in Cambridge, James Watson and Francis Crick were trying to build DNA with models. On the other side of the Atlantic, Linus Pauling of the California Institute of Technology had discovered the helix structure of proteins, and was doing his utmost to ensure that he would also go down in history as the man who discovered the structure of DNA.

DNA was known to consist of a chain of sugars, bases, and phosphates, but it was still not clear exactly how they were ranked within the DNA molecule. The researchers were equipped with the important information, discovered by Erwin Chargaff, that DNA contains equal amounts of the bases adenine and thymine, and of the bases cytosine and guanine.

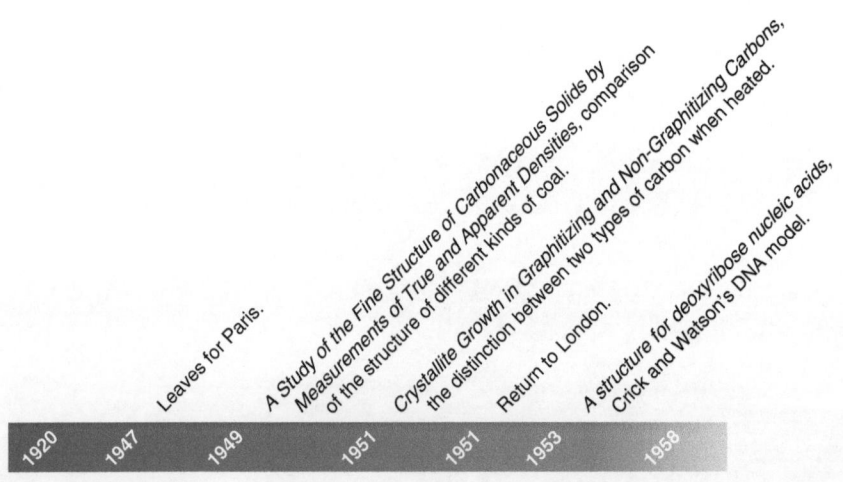

Rosalind Franklin

The race reached its climax in the first 2 months of 1953. In the first week of January, Pauling announced that he had found the structure of DNA, much to the dismay of his English rivals. They had already lost the race to find the protein structure, and now they seemed to have come in second again. When Watson saw the manuscript, however, his mind was quickly put at rest. Pauling's DNA model, a triple helix with the phosphate groups on the inside, was very similar to a model that Watson and Crick had developed – and discarded – 2 years previously.

At the end of January, there was a farewell seminar for Franklin at King's College, who was to leave for Birkbeck College later that year. Watson grasped the opportunity to talk to Franklin, but the conversation ended in a difference of opinion. Watson suggested that Franklin was not skilled enough to interpret the X-ray photographs correctly. Franklin felt insulted and showed Watson the door.

After Watson's hasty retreat from Franklin's room, he ran into Wilkins. Without Franklin's knowledge, Wilkins showed Watson one of her best photographs. Watson later wrote: "The instant I saw the picture, my mouth fell open and my pulse began to race."

A week later, Watson and Crick started to build a new model, this time with the phosphate groups on the inside of the molecule. They were still not sure whether to use a helix of two or three strands but, after seeing Franklin's photograph, Watson had set his mind on two, a double helix. They were also able to get hold of more of Franklin's measurements from an unpublished report of a recent visit by the Medical Research Council, which financed medical scientific research, to King's College. The report contained a table showing a number of crucial distances between the respective groups in a DNA molecule. From this, they could deduce that the two strands rotated in opposite directions, but it was not yet clear how they were held together.

At the end of February, Watson and Crick were able to put the final piece in the puzzle. By joining the two strands by hydrogen bridges between the adenine and thymine bases and between the cytosine and guanine bases, they had a perfect model that fitted exactly with the one-to-one relationship that Chargaff had discovered previously.

It is not clear if Franklin ever knew that Wilkins had shown the photograph to Watson. In their famous paper *A structure for deoxyribose nucleic acids*, Watson and Crick mentioned rather vaguely that they had "also been stimulated by a knowledge of the general nature of the unpublished experimental results and ideas" of Wilkins and Franklin. The sentence was formulated in such a way that Franklin would be unaware of the impact of her photograph. Watson, Crick, and Wilkins were awarded the Nobel Prize in 1962, but by then, Franklin had already been dead for 4 years.

Carbon

When Rosalind Franklin left the University of Cambridge in 1942, England was completely in the grip of the Second World War. There was great pressure – on women as well as men – to contribute to the war effort. Franklin did not want to

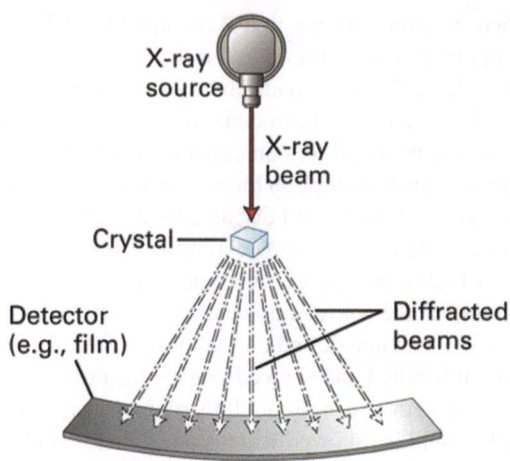

Rosalind Franklin made considerable use of X-ray diffraction in her research on both carbons and DNA. The dispersion of X-rays through solids with a crystal lattice is captured in a photograph. The pattern on the photograph provides information on the structure of the lattice

work in some stuffy office and hoped she could keep working in science in some way or another. She was therefore relieved to find a job at the British Coal Utilisation Research Association. That offered her the chance to combine business with pleasure. With England's war economy heavily dependent on coal, it was hardly surprising that the government pumped a lot of money into research into the "black gold."

Most of the coal in Great Britain came from fern-like plants, but the structure varies enormously. Franklin compared the structure of coal from England, Wales, and Ireland by measuring density and porosity. The dimensions of the pores are very important for coal's reactivity, as they determine how well it absorbs water and gas.

Franklin focused on the smallest pores, at molecular level. She compared "true density" with "apparent density." The latter was easy to measure by immersing the coal in a liquid that could not penetrate the finest pores. The true density was more difficult to determine, requiring a liquid or gas that penetrates the small pores without reacting with the coal.

To measure the true density, Franklin developed a method using helium. She then compared the results for true density with helium with the apparent densities, using hexane and benzene. She deduced that some pores were inaccessible for large molecules, like hexane and benzene, but were accessible for helium. This filtration property of coal had been known for some time, but had never been demonstrated as clearly with empirical measurements.

After the Second World War, Franklin started looking for work abroad, engaging the help of her friends. She wrote in a letter to Adrienne Weill, a fellow female Jewish scientist, asking her to let her know of job openings for "a physical chemist who knows very little physical chemistry, but quite a lot about the holes in coal." Not long afterward she received a fantastic offer to work at the *Laboratoire Central*

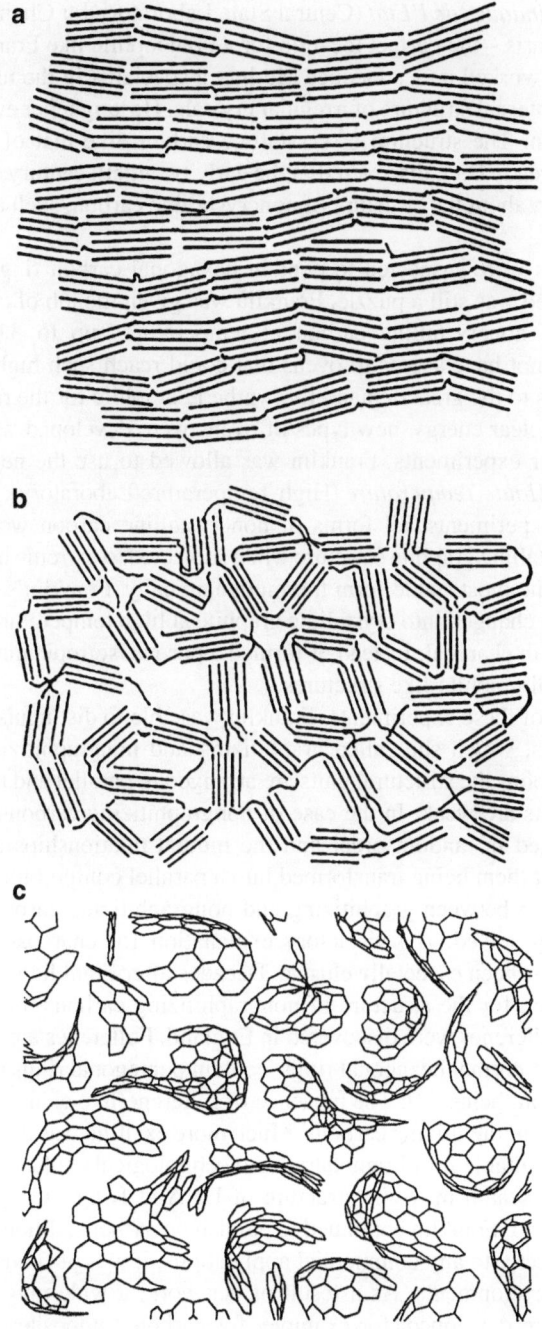

Franklin's model of graphitizing (a) and nongraphitizing (b) carbons, and (*below*) a recent model of nongraphitizing carbons, based on fullerene-like elements (c)

des Services Chimiques de l'Etat (Central State Laboratory for Chemical Services), in the heart of Paris – the perfect location for a Francophile like Franklin.

In Paris, she worked under Jacques Mering, a specialist in the use of X-rays to investigate the internal structure of irregular crystals. He taught her everything about X-ray diffraction. The structure of crystalline carbon, like that of diamonds and graphite, had already been discovered in the early twentieth century. Much less was known, however, about the structure of noncrystalline carbon, such as charcoal and coke.

Such carbons were suspected to contain hexagonal carbon rings, but how the rings were related was still a puzzle. Franklin was given the job of conducting heat experiments on noncrystalline carbon at temperatures up to 3,000°C. Earlier researchers did not have access to ovens that could reach such high temperatures. However, thanks to the great demand for synthetic graphite for the rapidly growing research into nuclear energy, new types of ovens were developed which made this possible. For her experiments, Franklin was allowed to use the new ovens at the *Laboratoire de Haute Temperature* (High Temperature Laboratory).

Before the experiments, all forms of noncrystalline carbon were expected to change to a crystalline graphite structure when subjected to extreme heat at 3,000°C, because this is the most stable form thermodynamically. Franklin's results showed that coke indeed changed into crystalline graphite at high temperatures, but that this did not occur with charcoal. Instead, it formed a porous isotropic material that contained only small graphite-like structures.

On the basis of these experiments, Franklin was able to distinguish between two types of carbons, which she called graphitizing and nongraphitizing carbons. In graphitizing carbons the structural units are arranged in parallel and the connections between the units are weak. In the case of nongraphitizing carbons, the structural units are arranged in random order and the mutual relationships are sufficiently strong to prevent them being transformed into a parallel configuration.

The distinction between graphitizing and nongraphitizing carbons has still not been completely solved. A satisfactory explanation for charcoal's resistance to graphitization has been especially elusive. Decades after Franklin's discovery, via a detour, the search for the structure of nongraphitizing carbons took a fascinating turn. In 1985, fullerenes were discovered in England. Fullerenes are carbons which, in addition to the normal hexagonal rings, contain pentagonal rings that prevent the carbon layers from being flat. The presence of fullerene-like elements may explain the stability of nongraphitizing carbons. Much more exciting was the fact that fullerenes are highly suitable for ultramodern nanotechnological applications.

During her research in to the structure of DNA at King's College, and in the years that followed, Franklin continued to publish on carbon. Although her research did not lead directly to any commercial applications, it was of fundamental importance to later developments. As a result of her work, a wide range of successful applications were developed, for example, for carbon composites, which consist partly of long carbon fibers. Carbon fibers are not very strong in themselves but, in combination with other materials, produce relatively light but ultra-strong fibers.

The discovery of graphitizing and nongraphitizing carbons was undoubtedly Franklin's main contribution to carbon sciences. Through a combination of clear insight, perseverance, and experimental skills, she succeeded in acquiring a crucial understanding of a substance that is not only very difficult to unravel, but also the most important element on Earth. Franklin's paper on the two types of carbon has become a classic of carbon literature, being cited 167 times in the last 10 years. Not bad for a paper that is half a century old.

References

Rosalind Franklin, 1949. 'A Study of the Fine Structure of Carbonaceous Solids by Measurements of True and Apparent Densities'. *Transactions of the Faraday Society* 45, 274–286.

Rosalind Franklin, 1951. 'Crystallite Growth in Graphitizing and Non-Graphitizing Carbons'. *Proceedings of the Royal Society of London.* Series A, Mathematical and Physical Sciences 209 (1097), 196–218.

Peter Harris, 2001. 'Rosalind Franklin's work on coal, carbon and graphite'. *Interdisciplinary Science Reviews* 26 (3), 204–210.

Brenda Maddox, 2003. *Rosalind Franklin: The Dark Lady of DNA*, Harper Collins, 380 pp.

George Gamow

Russian astrophysicist George Gamow is seen as the father of the Big Bang. Gamow was a creative thinker who felt quite at home taking a sidestep into another discipline. His contribution to cracking the genetic code is seen as "perhaps the last example of amateurism in scientific work on a grand scale."

R. Schils, *How James Watt Invented the Copier: Forgotten Inventions of Our Great Scientists*, DOI 10.1007/978-1-4614-0860-4_25,
© Springer Science+Business Media, LLC 2012

Big Bang

George Gamow grew up in Ukraine. In 1922, after a short sojourn at the University of Odessa, he went to the University of Petrograd, where his studies included the theory of relativity and quantum mechanics. Gamow's interest in astronomy was aroused by the lessons of Alexander Friedman, who concluded on the basis of Albert Einstein's general theory of relativity that the universe was not static, but either contracting or expanding.

Gamow planned to work with Friedman, but unfortunately the latter contracted pneumonia and died. Gamow was so disappointed that, in 1928, his teachers recommended him for a studentship at the University of Göttingen, at that time the center of quantum mechanics.

In Göttingen, Gamow used Schrödinger's wave equation to improve Ernest Rutherford's atomic model. He predicted the probability of the escape of alpha particles, consisting of two protons and two neutrons, from heavy unstable atoms. After spending some time in Copenhagen working with Niels Bohr, and then in Cambridge with Ernest Rutherford, Gamow returned to the Soviet Union where he specialized in the composition of atomic nuclei and radioactivity.

In 1933, Gamow was given permission to represent the Soviet Union at a major conference in Brussels. After giving his lecture, he did not return to his homeland, but fled to the USA. From 1935, he organized the annual Washington Conferences, a series of meetings on current research in nuclear physics.

During the eighth and final Washington Conference, Subrahmanyan Chandrasekhar and Louis Henrich attempted to explain the relative abundance of five different elements on Earth and in the universe. They came to the conclusion

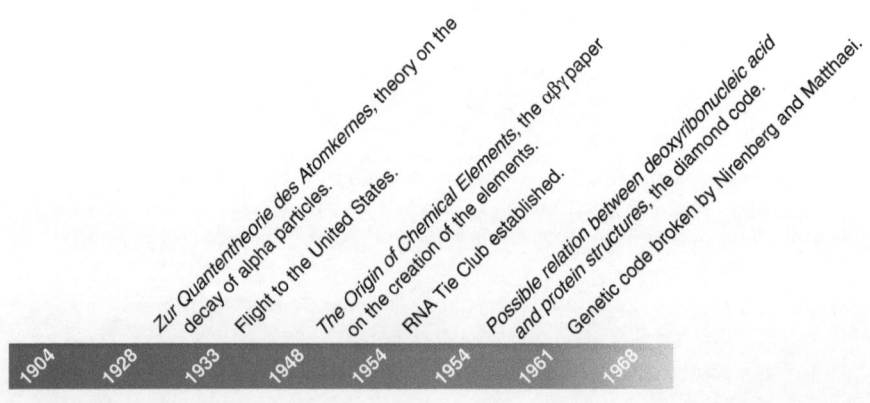

George Gamow

that the light and heavy elements could never have been formed under the same conditions, implying that when the elements were formed the universe could not have been in equilibrium.

After listening to Chandrasekhar and Henrich, Gamow devoted greater attention to the idea of an evolving universe. Between 1942 and 1946, he became convinced that the early universe must have been hot and compact. He reasoned that, as the universe cooled and expanded, neutrons must have separated into protons and electrons, and that the elements were then created by the capture of neutrons.

To calculate this scenario, a set of 300 differential equations had to be solved. This was not Gamow's favorite pastime but, fortunately, one of his PhD students, Ralph Alpher, found the job fascinating. Alpher's calculations led to the famous $\alpha\beta\gamma$ paper, named after its three authors – Alpher, Bethe, and Gamow. Hans Bethe did not actually help write the paper, but Gamow added his name without asking him so that he could use the catchy $\alpha\beta\gamma$ acronym.

The $\alpha\beta\gamma$ paper explained how all the elements could have been created simultaneously in the proportions we now find them from a hot, compact starting point. For hydrogen and helium, the two elements that together constitute 98% of the mass of the universe, Gamow and Alpher proved to have been correct. Most heavier elements, however, were not created in the first few minutes after the Big Bang, but billions of years later, in the hot cores of stars.

Gamow did not, incidentally, came up with the term Big Bang. It was first coined by British astronomer Fred Hoyle during a BBC radio broadcast in 1949. Gamow interpreted it as a negative reference to his model of the changing universe, despite Hoyle's claims to the contrary. Nevertheless, Gamow did not like the term and rarely – if ever – used it.

Genetic Code

In the spring of 1953, Francis Crick and James Watson decoded the structure of deoxyribonucleic acid (DNA). They discovered that DNA, the molecular basis of heredity, consists of two intertwined strands that run in opposite directions. Each strand is a long molecule comprising a string of sugars, phosphate groups, and one of the following four bases: adenine (A), thymine (T), cytosine (C), or guanine (G). The strands are connected by hydrogen bridges between adenine and thymine, and between cytosine and guanine. The burning question was soon raised: How is the information in DNA converted into the production of amino acids, the building blocks of proteins?

The first step in finding a solution came from an unexpected quarter. After reading the work of Watson and Crick, George Gamow wrote to them in the summer of 1953. He suggested that the base sequence in DNA might be the code for protein synthesis. As a physicist, Gamow's idea took the world of biology by storm. He had changed what had, until then, been seen as a chemical problem into purely a question of information storage and transfer. The underlying chemistry was of secondary importance.

Some members of the RNA Tie Club at Cambridge. From *left* to *right*: Francis Crick, Alex Rich, Leslie Orgel, and James Watson

Gamow had reduced the problem to the question: how can a language of four letters provide a code for 20 amino acids? It soon became clear that the four different bases had to be grouped in threes to make a unique code for each of the 20 amino acids possible. Groups of two only allow for 16 possibilities, while triplets provide 64, which is more than enough.

Gamow himself made the first proposal, what is known as the "diamond code." He thought that the protein synthesis occurred directly between the two strands of DNA. The four bases form a space in which an amino acid fits perfectly. Which acid that is depends on the bases at the four corner points, hence the name diamond. The bases on the top and bottom corners of the diamond lie on the same strand, separated by a single base. This base and its counterpart on the opposite strand constitute the left and right corners of the diamond. In essence, Gamow's was a three-letter code, as the left and right corners were complementary, so that only one of the two actually contained information.

Gamow's diamond was an overlapping code. Each base was part of three sequential triplets. For example, the base sequence ATCGAT consisted of the four triplets ATC, TCG, CGA, and GAT. Gamow came up with an original solution for the 64 possible triplets for only 20 amino acids. He suggested that the diamonds could, as it were, be rotated on both axes without that having any significance. If the TCA triplet were rotated on the horizontal axis, it would become ACT. Rotating it on the vertical axis would replace the middle base with its complement, making it TGA.

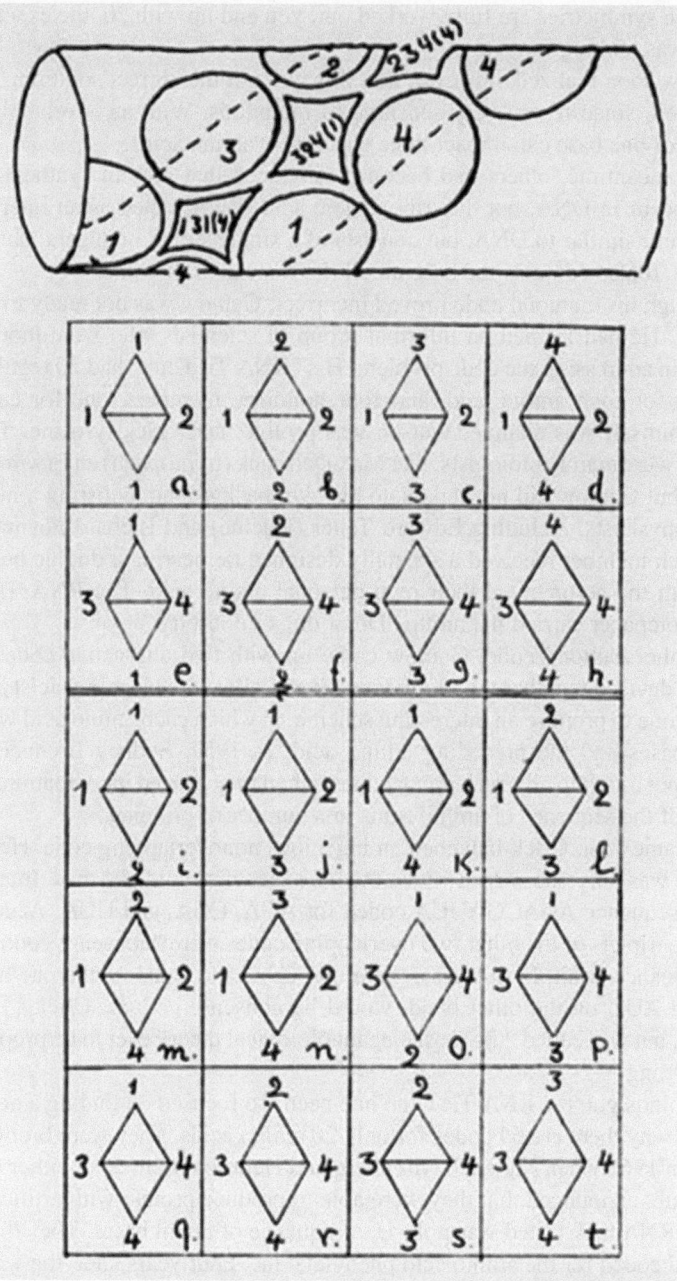

George Gamow's diamond code presumed that proteins were formed directly on the DNA. This drawing shows the bases, indicated with numbers *1–4*, and the 20 triplets (*a–t*)

If all these symmetries are fully worked out, you end up with 20, the exact number Gamow was looking for.

Gamow soon realized, however, that this was not the correct solution. This was just as well, since it was very sensitive to mutations. With an overlapping code, mutation of one base can impact three successive amino acids.

In the meantime, others had become convinced that protein synthesis did not directly occur in DNA, but that ribonucleic acid (RNA) acted as an intermediary. RNA is very similar to DNA, but consists of a single strand of sugars, phosphates, and bases. It also contains the base uracil (U) instead of thymine.

Although his diamond code proved incorrect, Gamow was not ready to throw in the towel. He had formed an informal group of scientists who were more or less involved in addressing the code problem. His "RNA Tie Club" had 20 regular members, one for each amino acid, and four honorary members, one for each base. Gamow himself was alanine, Watson was proline, and Crick tyrosine. The other members were mainly biologists, like Max Delbrück (tryptophan) en Erwin Chargaff (lysine), but Gamow did not repudiate his own background, enlisting a number of leading physicists, including Edward Teller (leucine) and Richard Feynman (glycine). Each member received a specially designed tie bearing a double helix and a tiepin with the acronym of their own personal amino acid. The RNA Tie Club's official notepaper carried the motto "Do or die, or don't try."

After the diamond code, Gamow came up with two alternative codes, one of which he devised together with Feynman. Even Teller, a nuclear physicist *pur sang*, took the time to propose an interesting scheme, in which each amino acid was coded by two bases and the preceding amino acid. In 1957, Sydney Brenner (valine) abruptly put a stop to all overlapping codes, when they proved incompatible with his analysis of the sequence of amino acids in a number of proteins.

That same year, Crick launched an ingenious nonoverlapping code. He claimed that there was only one way in which the base sequence could be read. Imagine that the base sequence AGACGAUUA coded for AGA, CGA, and UUA. According to Crick, the triplets of the other two overlapping codes were "nonsense codons," with no significance at all. In this case, therefore, GAC and GAU on the one hand, and ACG and AUU on the other hand, would be nonsense codons. Crick's code was incorrect, but was called "the most elegant biological theory ever to be proposed and proved wrong."

With hindsight, the RNA Tie Club had been too focused on finding a neat explanation of why there are 64 codes for only 20 amino acids. They were brought down to earth in 1961 when Marshall Nirenberg and Heinrich Matthaei, neither members of the Club, announced that they were able to produce protein with artificial RNA. The first RNA they tested was poly-U, a sequence of uracil bases. They discovered that UUU coded for the amino acid phenylalanine. Four years later, the whole coding problem was solved. Compared to the solutions proposed earlier, nature's solution seemed like a rather messy workaround. Some amino acids have only one codon, while others have four, and some even six. Although the real solution was less refined mathematically than his own idea, Gamow admitted that it had one great advantage: it was true.

References

George Gamow, 1954. 'Possible relation between deoxyribonucleic acid and protein structures'. *Nature* 173, 318.

Brian Hayes, 1998. 'The invention of the genetic code'. *American Scientist* 86 (1) 8–14.

Eamon Harper, 2001. 'In Appreciation George Gamow: Scientific Amateur and Polymath'. *Physics in Perspective* 3, 335–372.

James Watson, 2003. *Genes, Girls and Gamow*. Vintage Books, 336 pp.

Illustration Credits

Page 4
Deutsche Fotothek, Sächsischen Landesbibliothek, Staats- und
Universitätsbibliothek, Dresden.

Page 5
Photo Library, National Oceanic and Atmospheric Administration,
United States Department of Commerce.

Page 9
Picture Library, Museum of London.

Page 10
Guildhall Library, City of London.

Page 16
Edmond Halley, 1693. An Estimate of the Degrees of the Mortality of
Mankind, drawn from curious Tables of the Births and Funerals at the
City of Breslaw; with an Attempt to ascertain the Price of Annuities
upon Lives. Philosophical Transactions of the Royal Society of London
17: 596–610.

Page 17
Dirk van Ham.

Page 22
Daniel Bernoulli, 1954. Exposition of a New Theory on the
Measurement of Risk, Econometrica 22 (1), 23–36.

Page 23
Rijksmuseum, Amsterdam.

Page 28
Bibliothèque Nationale, Paris.

R. Schils, *How James Watt Invented the Copier: Forgotten Inventions
of Our Great Scientists*, DOI 10.1007/978-1-4614-0860-4,
© Springer Science+Business Media, LLC 2012

Page 29
Osher Map Library, Smith Center for Cartographic Education,
University of Southern Maine, Portland.

Page 34
Joseph Priestley, 1772. Impregnating Water with Fixed Air; In order to
communicate to it the peculiar Spirit and Virtues of Pyrmont water,
And other Mineral Waters of a similar nature. Johnson, 22 pp.

Page 35
Bottles and Bygones (http://www.bygonz.co.uk).

Page 39
Heriot-Watt University Archive, Records Management and
Museum Service, Edinburgh.

Page 41
Science & Society Picture Library, London.

Page 46
Naumann, Naturgeschichte der Vögel Mitteleuropas: Band iv,
Tafel 42 – gera, 1901.

Page 47
Natuur voor kinderen (http://natuur.ariena.com).

Page 51
The John Rylands University Library, The University of Manchester.

Page 53
David M. Hunt et al., 1995. 'The Chemistry of John Dalton's Color Blindness'.
Science 267 (5200) 984–988.

Page 59
Jean-François Champollion, 1824. Precis du Système Hieroglyphiques.
Impremerie Royal, 400 pp.

Page 67
Günther Klaus Judel, 2003. 'Die Geschichte von Liebigs Fleischextrakt'.
Spiegel der Forschung 20 (1), 6–17.

Page 73
Punch Magazine.

Page 74
Christian Feller et al., 2003. 'Charles Darwin, earthworms and the
natural sciences: various lessons from past to future'. Agriculture Ecosystems
& Environment 99, 29–49.

Page 80 and 81
History of the Atlantic Cable & Undersea Communications
(http://www.atlantic-cable.com).

Page 91 and 92
The Alexander Graham Bell Family Papers,
The Library of Congress, Washington.

Page 98 and 99
Hendrik Antoon Lorentz (voorzitter), 1926. Verslag Staatscommissie
Zuiderzee 1918–1926. Den Haag, 345 pp.

Page 106
United States Department of Energy.

Page 107
Svante Arrhenius, 1896. On the Influence of Carbonic Acid in the
Air upon the Temperature of the Ground'. Philosophical
Magazine 41, 237–276.

Page 114
Matthew Trainer, 2003. 'Kelvin and piezoelectricity'.
European Journal of Physics 24, 535–542.

Page 115
Der Piezoeffekt bei Kristallen (http://www.piezoeffekt.de).

Page 120
Fritz Wilhelm Winckel, 1931. 'Das Radio-Klavier von
Bechstein-Siemens-Nernst, Klangfarben auf Bestellung'. Die Umschau,
Illustrierte Wochenschrift über die Fortschritte in Wissenschaft und
Technik 35 (42), 840–843.

Page 121
Walther Nernst memorial (http://www.nernst.de).

Page 126
United States Patent and Trademark Office.

Page 127
Getty Images.

Page 134
Antweb (http://www.antweb.org), photographer: April Nobile.

Page 135
Harlow Shapley, 1920. 'Thermokinetics of Liometopum Apiculatum Mayr'.
Proceedings of the National Academy of Sciences 6 (4), 204–211.

Page 139
Erwin Schrödinger, 1944. What Is Life? The Physical Aspect of the
Living Cell. Cambridge University Press, 92 pp, and Österreichische
Zentralbibliothek für Physik.

Page 140
Irish Press, and Österreichische Zentralbibliothek für Physik.

Page 146
Alan Dunn, New Yorker Magazine.

Page 147
An Atlas of the Universe (http://www.atlasoftheuniverse.com).

Page 154
Department of Biology, University of Miami.

Page 155
Rosalind Franklin, 1951. 'Crystallite Growth in Graphitizing and
Non-Graphitizing Carbons'. Proceedings of the Royal Society of London.
Series A, Mathematical and Physical Sciences 209 (1097), 196–218,
and Peter Harris, University of Reading.

Page 161
Alexander Rich, and J.D. Watson Archives at Cold Springs Harbor Laboratory.

Page 162
George Gamow, 1954. 'Possible relation between deoxyribonucleic acid and
protein structures'. Nature 173, 318. Nature Publishing Group.